T0135671

Spyros Brezas

Investigation on the dissemination of unit watt in airborne sound and applications

Logos Verlag Berlin GmbH

λογος

Aachener Beiträge zur Akustik

Editors:
Prof. Dr. rer. nat. Michael Vorländer
Prof. Dr.-Ing. Janina Fels
Institute of Technical Acoustics
RWTH Aachen University
52056 Aachen
www.akustik.rwth-aachen.de

Bibliographic information published by the Deutsche Nationalbibliothek

The Deutsche Nationalbibliothek lists this publication in the Deutsche Nationalbibliografie; detailed bibliographic data are available in the Internet at http://dnb.d-nb.de .

D 82 (Diss. RWTH Aachen University, 2019)

ISBN 978-3-8325-4971-8
ISSN 2512-6008
Vol. 32

Logos Verlag Berlin GmbH
Comeniushof, Gubener Str. 47,
D-10243 Berlin
Tel.: +49 (0)30 / 42 85 10 90
Fax: +49 (0)30 / 42 85 10 92
http://www.logos-verlag.de

Spyros Brezas

Investigation on the dissemination of unit watt in airborne sound and applications

Investigation on the dissemination of unit watt in airborne sound and applications

Der Fakultät für Elektrotechnik und Informationstechnik der Rheinisch-Westfälischen Technischen Hochschule Aachen vorgelegte Dissertation zur Erlangung des akademischen Grades eines

Doktors der Ingenieurwissenschaften

Spyros, Brezas, M.Sc.

aus Larisa, Griechenland

Berichter:

Universitätsprofessor Dr. rer. nat. Michael Vorländer

Universitätsprofessor Dr.-Ing. Ennes Sarradj

Tag der mündlichen Prüfung: 12. Juni 2019

Investigation on the dissemination of unit watt in airborne sound and applications

Dissertation submitted to the Faculty of Electrical Engineering and Information Technology of the RWTH Aachen University for the award of the academic degree of

Doctor of Engineering

Spyros, Brezas, M.Sc.

from Larisa, Greece

Reviewers:

Prof. Dr. rer. nat. Michael Vorländer

Prof. Dr.-Ing. Ennes Sarradj

Date of the oral examination: 12. Juni 2019

This dissertation is available online on the website of the university library.

Στην κόρη μου Έλενα

και τη γυναίκα μου Ειρήνη

Contents

Abstract

Sound power is a widely applied quantity for the characterization of sound sources. Its determination is based on measurements of sound field quantities. Despite the state-of-the art measurement techniques, the sound power determination has some disadvantages. Most significant is the low frequency limitation, with different measurement methods leading to different results, which are expressed in broad frequency bands. A new method is proposed towards the establishment of traceability in airborne sound power. The realization of a primary source enables the free field sound power to be determined. The study investigates the dissemination process by which the sound power of a device under test can be referred to its free field sound power. In addition, the related uncertainty can be explicitly determined. The method of choice is the substitution method, which has been investigated both theoretically and experimentally. Apart from the well-established sound pressure measurements, the implementation of the substitution method also includes sound intensity measurements. The theoretical investigation focuses on the different positioning of the sources used in the substitution method, the substitution of sources of different radiation order and the existence of an impedance boundary. The sound power of aerodynamic reference sound sources has been examined since this type of source has been chosen to be the required transfer standard. For the measurements a specially designed scanning apparatus has been used. Sound power determination in calibration conditions and in situ has been performed. The required correction has been derived and successfully compared to an existing one. Sound pressure and sound intensity measurements at realistic environments have taken place and their sound power has been determined by applying the dissemination process. The sound power determination includes both narrow and broadband analysis along with a transparent uncertainty budget for the spectrum from 20 Hz to 10 kHz.

Introduction

When a source generates sound, it emits sound power causing sound to propagate through a medium (in airborne sound air, this is), which reaches the human ear, where it is detected as vibrations of the eardrum caused by the sound pressure fluctuations. In this model, sound power is the cause and sound pressure the effect. As sound power is closely related to the sound source, it is used to describe the source in terms of its acoustic characteristics.

Sound power cannot be directly measured, but it can be indirectly determined by measuring its effects: sound pressure and sound intensity. As it may be assumed, the effects do not only carry the wanted sound source information but are also influenced by additional conditions for the sound field. To understand the phenomena and to be able to distinguish their effects, some definitions follow on a number of basic acoustic concepts.

The sound power determination of any source includes measurements in an acoustic field. This field is the interference of three components: a) the field, which is radiated by the source itself and would exist in free field conditions, where sound propagates in absence of reflecting boundaries, b) the same type of field radiated by sources outside interest and c) the field related to physical phenomena such as reflection, scattering and diffraction caused by surrounding bodies and surfaces [Fah95]. The first component, is the one which describes the sound source of interest and can be further divided into three field conditions, which are related to the geometry of the source and the distance from it.

Closest to the source or adjacent to the vibrating surface is the hydrodynamic near field, which extends to a distance less than a wavelength [Bie03]. In this region, the pressure and particle velocity are nearly in quadrature ($\pi/2$ phase difference), there is fluid motion, but it is not directly related to sound propagation. Sound pressure measurements in this field cannot be considered indicative for the radiated sound power.

Next to the hydrodynamic field is the geometric near field [Bie03]. This extends to a larger area than the hydrodynamic. Sound pressure is generally not inversely proportional to the distance from the source, but relative minima and maxima occur. Although the sound pressure and particle velocity can be in phase or may have a small phase difference, there is no radial components from the source centre. Bies [Bie03] mentioned a study in his book to determine the sound power in the geometric near field but not with fully conclusive results [Bie93].

After the geometric near field, the far field starts and may extend to infinity. In this field the particle velocity and sound intensity have only radial components, and sound pressure and particle velocity are spatial angular [Fah95]. The far field exists when three criteria related to the distance approximately from the source, the wavelength and the characteristic source dimension are fulfilled [Bie03]. In the far field, the sound power can be determined along with the directivity. These two quantities can fully describe a sound source in terms of its radiated energy.

The existence of an ideally free field is not feasible, but the existence of free field conditions in confined spaces can be the case. The boundaries pose an impedance load to the source and change the effective radiation impedance. The changes in the radiation impedance have implications to the radiated sound power, especially when reflecting planes exist, where an increase of 3 dB in sound power is seen in case the source is positioned close to a reflecting plane, 6 dB near the junction of two reflecting surfaces and 9 dB in the junction of three surfaces [Bie03]. This means that the sound power that really exists in a realistic sound field is different from the sound power that would exist in a free field.

Sound power can be determined by a series of standardized measurements utilizing state-of-the-art measurement techniques. Each measurement technique is based on assumptions concerning the existing sound field, which are not fulfilled for the whole frequency spectrum between 20 Hz and 20 kHz, with large deviations at low frequencies. Figure I exhibits an example of the different sound power levels of the same source as a result of determination in different sound fields using different techniques. The current measurement methods determine the sound power that is really emitted, but it depends on the sound field and the measurement method.

This comes in contradiction to the characterization of the source based only to its characteristics. To overcome this contradiction, the present study proposes the characterization of the sound source based on its free field sound power, which is independent from the acoustic environment. Anechoic or hemianechoic rooms are qualified chambers, where free field occurs. In spite of determining sound power in such rooms, it is not possible for any source of interest either to be transported or to fit in such rooms.

The main concept of this study is to establish traceability for the unit watt in airborne sound. In plain words, this could enable the sound power of any source of interest located in any realistic environment to be referred to the free field sound power by a chain of measurements with pyramid structure. The actual sources, such as equipment or machinery, are located at the base of the pyramid. On top, a primary source of known free field sound power is located. The measurement chain is completed by the use of another source, which transfers the realistic sound power to the acoustic laboratory. If each measurement step is adequately defined, the related uncertainty can also be determined. This enables the sound power determination in metrology terms.

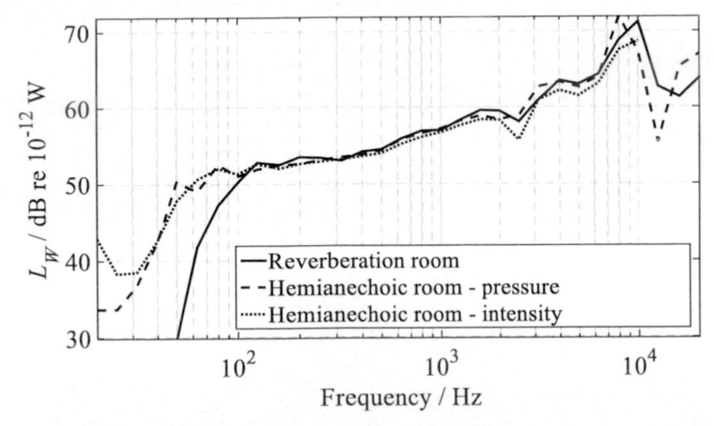

Figure I: Sound power level of a source determined in different sound fields using different measurement techniques.

In chapter 1, the definition of sound power is given along with a literature review about previous sound power research and the current limitations. The fundamental concept of the substitution method is explained in section 1.5, along with metrological terms. Chapter 2 presents a theoretical study of the substitution method, focusing on the order of the source and the existence of a reflecting plane. The calibration of transfer standards and the related uncertainty are discussed in chapter 3. The properties of transfer standards and the influence of environmental and operational conditions in their sound power are the subject of chapter 4. The determination of the sound power and the related uncertainty of sources under test in realistic environments are presented in chapter 5, including both sound pressure and sound intensity measurements.

1 Fundamentals of sound power

1.1 Sound power definition

According to ISO 80000 [ISO8000], the sound power P transported through a surface is the product of the sound pressure p and the component of the particle velocity u_n at a point on the surface in the direction normal to the surface, integrated over that surface. Sound power is expressed in watt and it relates to the rate per time at which airborne sound energy is radiated by a source.

The sound power level expressed in decibels, is defined by [ISO8000]:

$$L_W = 10\lg\left(\frac{P}{P_{\text{ref}}}\right)\text{dB} \tag{1.1}$$

where P_{ref} is the reference value (10^{-12} watt).

1.2 Use of sound power

The most common model in noise control problems is the source-path-receiver system. Three acoustic terms describe the action: emission, transmission and immision. Sound energy is initially emitted by a sound source, it is then transmitted through a propagation path and is ultimately immitted onto a receiver as it can be plainly explained by Figure 1.1. The source acoustic characteristics along with the transmission path are valuable information for the receiver's noise protection.

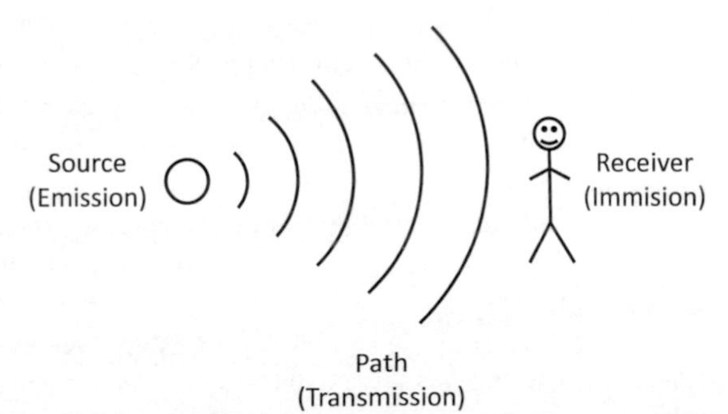

Figure 1.1: Sketch for the noise transmission from a source to a receiver.

Although the ear responses to sound pressure, the fact that this quantity is dependent on the distance from the source and on the acoustical environment constitutes it as an unsatisfactory descriptor of the emission. Instead, emission, which is related to the ability of the sound source to radiate sound at specific directions, is described by the sound power, which is not as greatly influenced by the surrounding environment [Fah95] and the directivity, which describes the variation of the sound radiation with direction [Ver06]. Sound power cannot be directly measured, but it may be determined by either sound pressure or sound intensity measurements.

Since sound power describes the total sound power of the sound source, it is a useful acoustic descriptor, which among others: enables the comparison of products of different suppliers, may verify if a product meets specific noise control requirements and can be used to predict the expected noise levels in diffuse or free field in case the directivity is additionally known for the latter [Bie03]. The sound power is a commonly used quantity in legislation or specifications related to the sound energy emitted by sound sources. As examples, the European Union Outdoor Directive [Dir14] requires the determination and declaration of the sound power of equipment for use outdoors, the Machinery Directive [Dir42] relates the sound power to requirements for safety issues and the Energy Labelling Directive [Dir30] uses the sound power for standard product information. For the enhanced product data evaluation by a consumer, the quote of the sound power information must be comprehensively declared [Gue16]. The sound power, apart from a significant acoustic quantity, is a consumer and trade tool, which must be kept updated to the increasing demand for lower noise levels in modern daily life. For this, improvements must be applied to many sound power aspects. For example, the determination of the related uncertainty is of great importance [Car07] along with a new noise labelling [Car16].

The quantity of sound power is currently used in more than fifty ISO standards either as the main topic (the term sound power included in the title) or as a supplementary topic (the term mentioned in the main text) covering different types of sources (e.g. noise sources, earth moving machinery, industrial fans, reciprocating internal combustion engines etc.). The present study focuses on the sound power determination of noise sources described by the ISO standards belonging to acoustics field. There are seven standards determining sound power by sound pressure measurements: ISO 3714 [ISO3741], ISO 3743-1 [ISO37431], ISO 3743-2 [ISO37432], ISO 3744 [ISO3744], ISO 3745 [ISO3745], ISO 3746 [ISO3746] and ISO 3747 [ISO3747]. Three standards use sound intensity as the measurable quantity for sound power determination: ISO 9614-1 [ISO96141], ISO 9614-2 [ISO96142] and ISO 9614-3 [ISO96143]. Additionally, ISO 6926 [ISO6926] is also included, because it describes the requirements of reference sound sources used for the sound power determination.

1.3 State-of-the-art in sound power measurements

Sound power has been for a long time a field for wide acoustic research. In this paragraph, a literature overview is given to end up to the up-to-date sound power measurement status. The overview mainly focuses on the topics, which will be further discussed in the next chapters. These are: theoretical sound power studies, the use of reference sound sources in sound power measurements, the use of sound pressure and sound intensity to the sound power measurements and the related uncertainties.

The direct determination of sound power after sound pressure or sound intensity measurements or after the application of the substitution (mentioned also as comparison) method has been the topic of many studies. Lubman presented the errors in sound power determination using both the direct and the comparison method in reverberant environments [Lub74]. Hübner [Hü731] discussed the errors concerning the direct sound power determination in free field conditions in terms of different measurement surfaces and different source positioning within the measurement surface. The determination of sound power above a reflecting plane in free field conditions was analyzed by Holmer [Hol77]. The influence of a reflecting plane in sound intensity measurements was discussed by Pope ([Pop86], [Pop89]). Russell [Rus80] calculated the error propagation equations for various sound power measurement techniques, including the substitution method in reverberation rooms. Bies investigated the near field in sound power determination based on sound pressure measurements [Bie93]. The near field error was also the subject of An-tze [Ant08] paper. Probst [Pro89] argued about the angle error concerning the sound determination over an enveloping surface including the mirror source modelling. The differences between sound power levels by measurements at different acoustic fields and their consequences to acoustic measurements were examined by Vorländer [Vor952]. Free field sound power measurements are performed in qualified anechoic or hemianechoic rooms. The measurement bandwidth is an influential factor for the qualification procedure and subsequently for the sound power determination in narrow bands as it was shown by Wittstock [Wit042] and Cunefare [Cun06].

The environmental conditions affect the radiated sound power and therefore many studies have dealt with the related corrections. The required acoustic environment correction for the determination of the sound power level was examined by Hübner [Hü771], who also presented a first approach for the comparison of sound power under reference meteorological conditions [Hü801]. Hübner discussed about the relation of the sound power to normalized environmental conditions [Hü992] and about the influence of environmental conditions (static pressure) in the sound generation by aerodynamic sources [Hü772]. The same author suggested a correction for the effect of changing atmospheric pressure to sound generation by fans [Hü981]. A research on the calibration procedure of reference sound sources was performed by Tsuei [Tsu14], focusing on the environmental correction factors.

The errors in sound power determination are related to uncertainties. An attempt to determine the uncertainty of sound power by the measurement reproducibility in hemianechoic environments was made by Hanes [Han92]. Finke [Fin93] presented a study on the uncertainty components during the sound power determination according to ISO 3745 [ISO3745]. The need of clear statements of the uncertainty in sound power measurements was highlighted by Higginson [Hig93]. Jacobsen provided an overview on the sources of uncertainty in sound power measurements using sound intensity ([Jac97], [Jac07]). The differences between the true and the determined sound power value using the comparison method were investigated by Sehrndt [Seh96]. The sound power level and its uncertainty by varying the source acoustic parameters in a hemianechoic room were examined by Simmons [Sim04]. Loyau [Loy07] explored the uncertainty related to the number of microphones, the angle and the air impedance during sound power determination using sound pressure measurements over a hemisphere. A work was undertaken by Caligiuri [Cal07], who focused on the main uncertainty components for the sound power determination in various cases. A study for the uncertainty of sound power emitted by pure-tone or narrow band sources in a reverberation room was presented by Jacobsen [Jac09].

An important aspect of sound power determination is the use of reference sound sources. The variations of the sound power of a sound source with the environment it is placed in, were studied by Ballagh [Bal82] using both a monopole and a dipole. The use of reference sound sources in the determination of the sound power and the related accuracy was presented by Jonasson ([Jon86], [Jon88]). Tachibana studied the differences in sound power in different boundary conditions by measuring sound intensity [Tac89]. The need for narrow band sound power determination was discussed by Hickling [Hic90]. The determination of sound power in reverberation rooms focusing on the low frequencies was studied both theoretically and experimentally by Agerkvist [Age93]. The sound power determination in factory halls by applying the substitution method was discussed by Probst [Pro93]. A model to study the sound power level deviations of reference sound sources in low and high frequencies was presented by Campanella [Cam96]. A very comprehensive study about the calibration of reference sound sources in free and reverberant field was performed by Vorländer [Vor951]. An experimental study dealt with the influence of a reflecting plane in an anechoic room to the hemifree field sound power of a reference sound source [Yam14].

Measurement automation was on the focus of some studies. Firstly, the measurement of sound power using a moving microphone was applied [Brü76]. An automated system for sound power measurements was presented by Yanagisawa [Yan82]. The use of the same kind of system for the measurement of sound intensity for the further sound power determination was designed by Hickling [Hic97].

Apart from the well established aerodynamic reference sound sources, alternatives have been suggested. A proposal for an impulse reference sound source was made by Tachibana [Tac96]. The use of a calibrated dodecahedron loudspeaker as a reference sound source was proposed by Dragonetti [Dra11]. Since the aerodynamic reference sound sources generate wind, it is common for windscreens to be used. The effects of the windscreens in sound pressure measurements were presented by Hessler [Hes08] and in intensity probes by Jacobsen [Jac94]. A numerical investigation on the use of windscreens was performed by Juhl [Juh06].

Influences on sound power determination have also been studied theoretically. The reflection of a sound wave upon a reflecting plane has been the topic of many studies. An initial approach was made by Ingard [Ing51]. A correction for the previous study was proposed by Thomasson [Tho76]. A simulation for the sound power determination based on sound intensity utilizing a computer program was performed [Lav92]. Another simulation on the sound power determination was provided by Wu [Wu86], where a theoretical model was developed to compare the sound power after a finite number of sound intensity measurements and the actual power that passes through the same surface. The sound intensity of two interfering monopoles in the near field was the scope of the study of Krishnappa [Kri83]. Taking this work a step further, the same author provided both theoretical and experimental results on near field sound intensity of a monopole over a reflecting boundary [Kri87]. The number of samples on the measuring surface was discussed by Tohyama [Toh87] for the free field method. Continuing his previous research, Tohyama [Toh88] considered different types of finite-sized sources for sound power determination. The errors due to the finite distance between the two microphones of the intensity probe were calculated based on theoretical expressions given by Shirahatti [Shi88].

Mechel discussed the propagation of a spherical wave above a reflecting plane [Mec891]. Tohyama [Toh90] also investigated the errors due to the determination of the averaged mean sound pressure over the measurement surface. A theoretical study about the errors in sound intensity measurement in case of a dipole source was discussed by Shirahatti [Shi92]. The mathematical background for the sound intensity scanning method was calculated by Paterson [Pat93]. The effects of sound pressure interference were discussed by Corrêa [Cor94]. For the propagation of sound above a reflecting plane, a model based on the complex image theory was also presented [LiY96]. A floor reflection model was used to investigate the differences on the sound power level of a reference sound source related to the distance between the source acoustic centre and the microphone [Cor96]. The sound propagation near an impedance plane concerning a dipole was discussed by Li [LiK97]. A sound source model above an infinitely large floor was designed to study the errors in sound power determination based on both sound pressure and sound intensity measurements [Suz07]. The reflection from a rough surface was considered for the theoretical calculation of the sound intensity of a monopole source [Max12]. The effect of a reflecting plane on the calibration of a reference sound source was experimentally and theoretically investigated by Yamada [Yam15]. The effect of a finite size reflecting boundary on the determination of the sound power in an anechoic room was also analyzed by Zhong [Zho18].

1.4 Limitations of current methods

The sound power determination of airborne sound can be performed by sound pressure, sound intensity and vibration measurements. The guidelines for the selection of the appropriate method are described by ISO 3740 [ISO3740]. The main parametes of the selection are the type of the sound source, the measurement environment and the desired accuracy. The present study focuses on the sound power determination based on sound pressure and sound intensity measurements as it has already been described in section 1.2. Table 1.1 summarizes the state-of-the-art sound power determination applications according to the up-to-date ISO standards [ISO3740].

Based on the table, sound power can be either directly determined by sound pressure or sound intensity measurements or determined by the substitution (comparison) method by sound pressure measurements using a reference sound source. There is no procedure to refer the sound power of a source to the sound power of a primary source. Although at each standard the related uncertainties are provided, an uncertainty budget to explain in detail the sources of uncertainty at each step of the sound power determination is lacking. Only broadband frequency analysis may be used from 100 Hz for sound pressure and 50 Hz for sound intensity measurements. Below the low frequency limits, there are deviations among the different sound power determination procedures. A scanning mechanism to measure sound pressure or sound intensity of a stationary source over a fully covered surface does not exist. Any corrections related to the reference sound sources assume their dipole behaviour, which has not been explicitly studied. To overcome the above limitations is the motivation and the starting point of this thesis.

Before the discussion of the theoretical and experimental results, an introduction to the metrological term of traceability is given in the next section. The substitution method is also explained, since it is the basis for the calculations presented in the study.

Table 1.1: Up-to-date standard specifications for sound power determination.

3741: 2010	**3743-1: 2010**	**3743-2: 2018**	**3744: 2010**	**3745: 2012**	**3746: 2010**	**3747: 2010**
Sound pressure						
Acoustic environment						
Reverberation room	Hard-walled room	Special reverberation room	Essentially free field over reflecting plane	Anechoic or hemianechoic room	In situ over reflecting plane	Essentially reverberant field in situ
Precision grade						
Precision (1)	Engineering (2)			Precision (1)	Survey (3)	
Character of source sound						
Steady, broadband, narrow band or discrete frequency	Steady, broadband, narrow band or discrete frequency		Any			Steady, broadband, narrow band or discrete frequency
Source volume						
Volume no more than 2% of test room volume	Volume no more than 1% of test room volume		No restrictions	Measurement radius more than double characteristic dimension	No restrictions	
Frequency analysis						
1/3 oct	1/1 oct		1/3 oct		1/1 oct	
Lowest frequency						
100 Hz	125 Hz		100 Hz		125 Hz	
Sound power determination method						
Direct Substitution	Substitution	Direct Substitution	Direct			Substitution

(Table continues to next page for sound intensity measurements.)

9614-1: 1993	**9614-2: 1996**	**9614-3: 2002**
Sound intensity		
Acoustic environment		
No requirement		
Precision grade		
Precision, engineering, survey (1,2,3)	Engineering, survey (2,3)	Precision (1)
Character of source sound		
Steady, broadband, narrow band or discrete frequency		
Source volume		
No restrictions		
Frequency analysis		
1/3 oct		
Lowest frequency		
50 Hz		
Sound power determination method		
Direct		

1.5 Traceability of sound power unit

A new approach has been proposed for the sound power determination ([Wit13], [Wit14]). According to this approach, the in situ sound power of a sound source may be referred to the free field sound power of the same source by establishing traceability.

By definition [JCG12], traceability is the property of a measurement result when it can be related to a reference through a documented unbroken chain of calibrations. Figure 1.2 shows the calibration chain for the proposed traceability of the unit watt in airborne sound. The free field sound power of a primary source under calibration conditions (top) is disseminated by a transfer source (middle) and used for the determination of the in situ sound power of real sources (bottom) using the substitution method. This way, the free field sound power of the real sources is expected to be determined. The application of the substitution method at each dissemination step ensures except from the sound power, the determination of the related uncertainty as well, which is also a prerequisite for the establishment of traceability.

The following sources may be used as an introduction to metrology concepts: [Tay94], [Gol10], [Und18], [Bir03], [Bel01], [Str18].

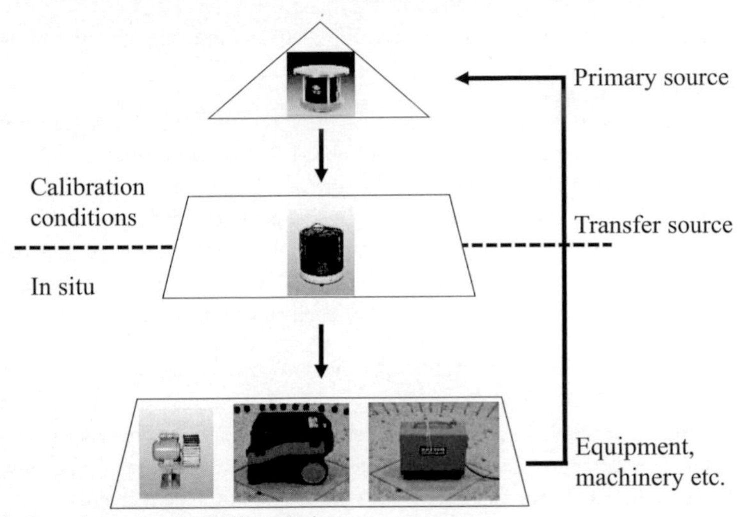

Figure 1.2: Sound power traceability chain.

The substitution method relates the known sound power level of a source to the unknown sound power level of another source by also calculating the difference between the surface and time averaged sound pressure levels of the two sources. It is described by:

$$L_{W,\,\text{unknown}} = L_{W,\,\text{known}} + \overline{L_{p,\,\text{unknown}}} - \overline{L_{p,\,\text{known}}} \tag{1.2}$$

where L_W is the sound power level and $\overline{L_p}$ the surface and time averaged sound pressure level [Ver06].

To avoid low frequency effects such as near field and room modes, the present contribution also investigates the application of sound intensity level L_I, to the substitution method. Eq. (1.2) is accordingly modified to:

$$L_{W,\,\text{unknown}} = L_{W,\,\text{known}} + \overline{L_{I,\,\text{unknown}}} - \overline{L_{I,\,\text{known}}} \tag{1.3}$$

The related uncertainties can be generally described by:

$$u^2\left(L_{W,\,\text{unknown}}\right) = u^2\left(L_{W,\,\text{known}}\right) + u^2\left(\overline{L_{p,\,\text{unknown}}} - \overline{L_{p,\,\text{known}}}\right) \tag{1.4}$$

$$u^2\left(L_{W,\,\text{unknown}}\right) = u^2\left(L_{W,\,\text{known}}\right) + u^2\left(\overline{L_{I,\,\text{unknown}}} - \overline{L_{I,\,\text{known}}}\right) \tag{1.5}$$

where u^2 is the variance of the probability distribution of the quantity in parenthesis [JCG08].

Sound power traceability aims to improve the sound power determination deficiencies as described previously and shown in Table 1.1. The application of the substitution method is expected to eliminate low frequency limitations and thus, to expand the usable frequency range below 50 Hz. Finer frequency resolution would enable the sound power determination of tonal sources. A specially designed apparatus would provide more robust surface and time averaged levels. A combined uncertainty can be determined by the partial uncertainties of each substitution step.

A starting point for the determination of sound power using the substitution method by including both sound pressure and sound intensity is the investigation of the procedure in theory. A number of models were implemented in order to examine the factors that influence the substitution method and to which extend. The description of the modelling and the related results are presented in the next chapter.

2 Theoretical considerations for the substitution method

The theoretical determination of sound power has been an intriguing topic for many acoustic studies. These studies have mainly focused on the direct sound power determination, and few have been related to the substitution method, providing the need for further investigation. In this chapter various influential factors of the substitution method are examined, in order to provide a theoretical background for the evaluation and interpretation of the substitution method results based on both sound pressure and sound intensity measurements as presented in chapter 5.

The following paragraphs describe the theoretical implementation of the substitution method focusing on various influential factors. Different source orders were considered (monopoles and dipoles) and the substitution method was applied using both sound pressure and sound intensity. The substitution includes the comparison of sources of the same order (e.g. monopole-monopole and dipole-dipole) and of different order (e.g. monopole-dipole). Calculations were performed for free field conditions and for the presence of reflecting planes. For the latter, both plane and spherical wave approach were considered.

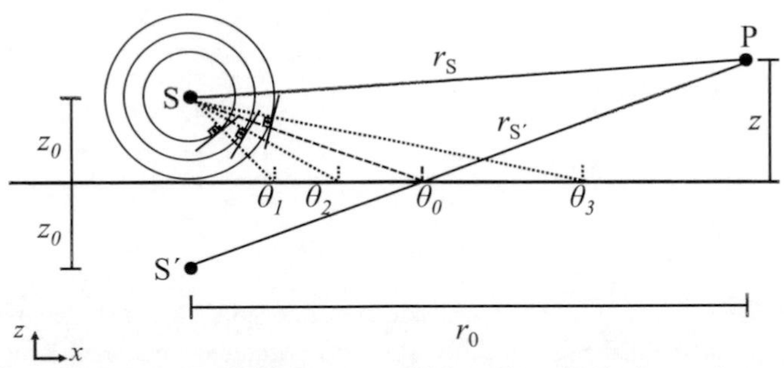

Figure 2.1: Geometry of a point source, which emits spherical sound waves and a receiver for sound emission above a reflecting plane.

2.1 Reflection over impedance plane

The substitution method is applied in realistic acoustic environments, which include reflections from surfaces. A simple example is the highly reflecting floor of a hemianechoic room. For the influence of such surfaces on the theoretical calculation of the sound pressure and sound intensity over a surface, two methods were considered. Firstly, the Sommerfeld [Som09] solution was implemented and secondly, the mirror source model [Vor08]. In both methods only specular reflections were considered since the investigation was more focused on lower frequencies.

Figure 2.1 shows a point source S emitting spherical waves above a reflecting plane while the direct and reflected sound is detected by a receiver P. Mechel [Mec892] provided a comprehensive study for the derivation of the sound pressure at the receiver point, which was used as the basis for the analysis in this chapter. The positioning of the source and receiver was described by spherical coordinates. An example of a spherical coordinate system is shown in Figure 2.2.

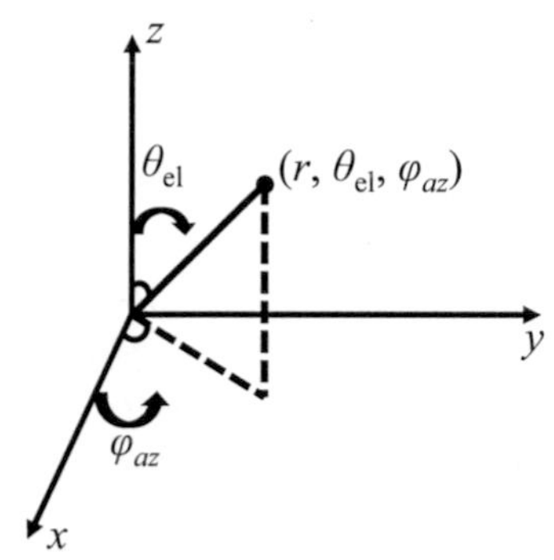

Figure 2.2: Spherical coordinate system.

According to Figure 2.2, the position of the source and the receiver may be described in spherical coordinates as:

$$\begin{aligned} x &= r \sin\theta_{el} \cos\varphi_{az} \\ y &= r \sin\theta_{el} \sin\varphi_{az} \\ z &= r \cos\theta_{el} \end{aligned} \tag{2.1}$$

where r is the distance from the coordinate system origin, θ_{el} is the elevation angle and φ_{az} is the azimuthal angle (the elevation and azimuthal angle increments follow the corresponding arrow notations of Figure 2.2).

The sound pressure at the receiver point consists of the direct and the reflected part and is given by:

$$p(x,y,z) = p_{dir} + p_{refl} = j\rho c k q \frac{e^{-jkr_S}}{4\pi r_S} + p_{refl} \tag{2.2}$$

where ρ is the air density in kg/m^3, c the sound speed in air in m/s, k the wavenumber in m^{-1} and q the effective complex source strength in m^3/s [Fah98].

As it can be seen in Figure 2.1, the sound reflections are strongly dependent on the incidence angle on the reflecting plane. In order to provide an exact solution, Mechel

[Mec892] suggested the decomposition of the spherical wave into an infinite sum of plane waves. Thus, the reflected sound is described by:

$$p_{\text{refl, spherical}} = \frac{\rho c k^2 q}{4\pi} \int_{\Gamma_\theta} J_0\left(kr\sin\theta\right) e^{-jk\left(z+z_0\right)\cos\theta} R\left(\theta\right) \sin\theta \, d\theta \tag{2.3}$$

where J_0 is the Bessel function of zeroth order. For the calculation of the Sommerfeld integral in Eq.(2.3) complex angles must be taken into consideration to account for all possible directions of the propagation vector. As it can be seen in Eq.(2.3), the angle dependent reflection factor $R(\theta)$ and the phase shift related to the path followed by each reflected plane wave from the source to the receiver are taken under consideration. The angle dependent reflection factor is given by:

$$R\left(\theta\right) = \frac{\xi\cos\theta - 1}{\xi\cos\theta + 1} \tag{2.4}$$

where ξ is the specific acoustic impedance of the reflecting plane normalized by the air characteristic impedance ρc .

Suh [Suh99] provided a well-established numerical integration method for Eq.(2.3). According to this approach, the Sommerfeld integral is divided into a definite integral for the interval $[0, \pi/2)$ and an improper integral for the interval $(\pi/2 + j0, \pi/2 + j\infty]$. Suh [Suh99] showed that the improper integral converges except grazing incidence and that the angle step $\Delta\theta$, which replaces $d\theta$, must be as small as possible for better accuracy.

For an approximation of Eq.(2.3) the mirror source model is usually used. The simplification of the configuration of Figure 2.1 is shown in Figure 2.3, where the source sound waves can be considered to be plane instead of spherical. In this model, the principles of geometrical acoustics are followed. Similarly to optics, in geometrical acoustics the sound waves are dealt as rays [Vor08]. The mirror source model is widely used in room acoustics, especially after Allen [All79] showed that it can provide results equivalent to those after solving the Helmholtz equation for the case of rectangular rooms. The approximated form of the reflected sound becomes:

$$p_{\text{refl, plane}} = j\rho c k q \frac{e^{-jkr_{S'}}}{4\pi r_{S'}} R\left(\theta_0\right) \tag{2.5}$$

The approximation of $p_{\text{refl, spherical}}$ to $p_{\text{refl, plane}}$ may be the cause of errors especially in the near field, because it violates the wave equation [Mec892]. Then three main error sources may be identified when: the sum of the source and the detector heights $(z + z_0)$ is not large enough compared to the wavelength, the reflection factor R

exhibits strong angular variation at the specular reflection angle θ_0 and the specular reflection angle θ_0 is close to grazing incidence. Usually, the Sommerfeld integral is used in room acoustics for the error calculation, while the mirror source model is applied for the main sound field calculations ([Suh99], [MAr14]).

An influential factor of the substitution method is the source positioning and its distance to the receiver. Additionally, near field effects also influence the substitution method. The theoretical calculations of the substitution method were performed including both the Sommerfeld integral and the mirror source approach.

For the purposes of the present study, the substitution method has also been applied using sound intensity apart from sound pressure. The calculation of the sound intensity based on the sound pressure follows.

The sound intensity vector of the source S in the direction of the radial component r_S can be calculated according to [Fah98]:

$$I_\mathrm{r} = \frac{1}{2}\mathrm{Re}\left[p\left(r_\mathrm{S}\right)\cdot u_\mathrm{r}^*\left(r_\mathrm{S}\right)\right] \tag{2.6}$$

with u_r^* being the complex conjugate of the radial particle velocity in m/s, which is related to the sound pressure according to [Fah98]:

$$u_\mathrm{r}\left(r_\mathrm{S}\right) = -\frac{1}{j\omega_\mathrm{ang}\,\rho}\frac{\partial p\left(r_\mathrm{S}\right)}{\partial r_\mathrm{S}} \tag{2.7}$$

where ω_ang is the angular velocity in rad/s.

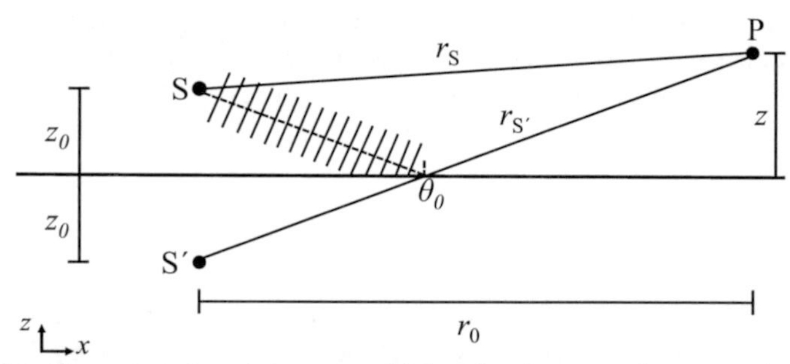

Figure 2.3: Geometry of a point source, which emits plane sound waves and a receiver for sound emission above a reflecting plane.

The estimation of the gradient can be approximated [Ver06] and Eq.(2.7) becomes:

$$u_{\mathrm{r}}\left(r_{\mathrm{S}}\right)=-\frac{1}{j\omega_{\mathrm{ang}}\rho}\frac{p\left(r_{\mathrm{S}}+\Delta r/2\right)-p\left(r_{\mathrm{S}}-\Delta r/2\right)}{\Delta r} \tag{2.8}$$

where Δr is the distance in m of the positions at which the sound pressure is determined and must be much smaller than the wavelength of interest. The distance Δr is equivalent to the length of the microphone spacer of the intensity probe used for sound intensity measurements. For the simplification of the sound intensity calculation, all distances between each source and each receiver were expressed in relation to the axis origin, because this ensures the calculation of the intensity normal to the measurement surface component as indicated by Eqs.(2.7) and (2.8).

In general, the distance between two points, positioned at (x_1, y_1, z_1) and (x_2, y_2, z_2), may be calculated by:

$$\Delta=\sqrt{\left(x_2-x_1\right)^2+\left(y_2-y_1\right)^2+\left(z_2-z_1\right)^2} \tag{2.9}$$

As an example, the distance r_{S} between the source $\mathrm{S}\left(z_0, 0, 0\right)$ and the receiver $\mathrm{P}\left(r, \theta_{\mathrm{el}}, \varphi_{\mathrm{az}}\right)$ in Figure 2.1 is:

$$r_{\mathrm{S}}=\sqrt{\left(-r\sin\theta_{\mathrm{el}}\cos\varphi_{\mathrm{az}}\right)^2+\left(-r\sin\theta_{\mathrm{el}}\sin\varphi_{\mathrm{az}}\right)^2+\left(z_0\cos 0-r\cos\theta_{\mathrm{el}}\right)^2} \tag{2.10}$$

which ultimately provides:

$$r_{\mathrm{S}}=r\sqrt{z_0^2/r^2+1-2z_0/r\sin\theta_{\mathrm{el}}\cos\varphi_{\mathrm{az}}} \tag{2.11}$$

Using Eq.(2.11) the direct sound may be expressed as:

$$p_{\mathrm{dir}}=j\rho c k q\frac{e^{-jkr\sqrt{z_0^2/r^2+1-2z_0/r\sin\theta_{\mathrm{el}}\cos\varphi_{\mathrm{az}}}}}{4\pi r\sqrt{z_0^2/r^2+1-2z_0/r\sin\theta_{\mathrm{el}}\cos\varphi_{\mathrm{az}}}} \tag{2.12}$$

According to the analysis of Suh [Suh99], Eq.(2.3) is expressed as:

$$p_{\text{refl, spherical}} = \frac{j\rho c k^2 q}{4\pi} \left\{ \begin{array}{l} j\int_0^{\frac{\pi}{2}} J_0\left[kr\left(\sin\theta_{\text{el}}\right)\sin\Theta_1\right] \\ e^{-jkr\left(\cos\theta_{\text{el}}+z_0/r\right)\cos\Theta_1} \times \dfrac{\xi\cos\Theta_1 - 1}{\xi\cos\Theta_1 + 1}\sin\Theta_1\, d\Theta_1 \\ -\int_0^{\infty} J_0\left[kr\left(\sin\theta_{\text{el}}\right)\cosh\Theta_2\right] \\ e^{-kr\left(\cos\theta_{\text{el}}+z_0/r\right)\sinh\Theta_2} \times \dfrac{-j+\xi\sinh\Theta_2}{j+\xi\sinh\Theta_2}\cosh\Theta_2\, d\Theta_2 \end{array} \right\} \tag{2.13}$$

where Θ_1 and Θ_2 are the real and imaginary reflection angles, respectively.

The reflected sound for the mirror source S' positioned at $\left(z_0, \pi, 0\right)$ and by applying Eq.(2.10) is expressed as:

$$p_{\text{refl, plane}} = \\ j\rho c k q \frac{e^{-jkr\sqrt{\left(z_0/r\right)^2+1+2z_0/r\cos\theta_{\text{el}}}}}{4\pi r\sqrt{\left(z_0/r\right)^2+1+2z_0/r\cos\theta_{\text{el}}}} \times \frac{-1+\dfrac{\xi}{\sqrt{1+\dfrac{\sin^2\theta_{\text{el}}}{\left(\cos\theta_{\text{el}}+z_0/r\right)^2}}}}{1+\dfrac{\xi}{\sqrt{1+\dfrac{\sin^2\theta_{\text{el}}}{\left(\cos\theta_{\text{el}}+z_0/r\right)^2}}}} \tag{2.14}$$

The same analysis may be applied to the equations for the particle velocity of the reflected sound for both spherical and plane wave approach.

2.2 Measurement surface

The substitution method requires the determination of the surface and time averaged sound pressure or sound intensity of both sources. A grid of receivers may sample the surface over which the sound power is to be determined. This comes in contrast to room acoustics calculations, where a pair of source and a receiver is usually considered. The number of receivers increases the computational time, especially in the case of the Sommerfeld integral evaluation [MAB16].

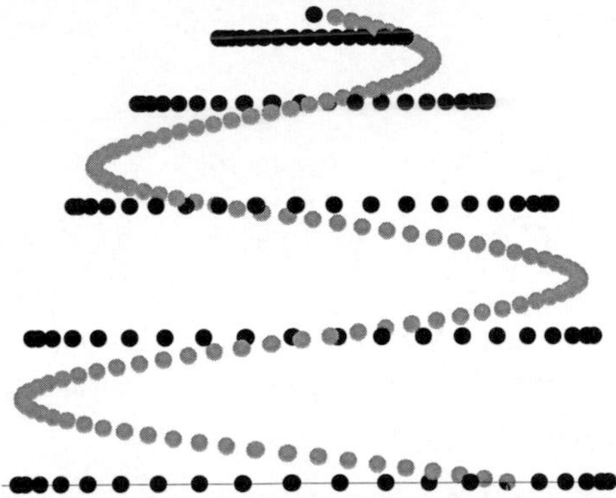

Figure 2.4: Hemispherical receivers grid for coaxial circular (black) and helicoid (grey) positioning.

The measurement surface was chosen to be spherical for free field considerations and hemispherical for the case of hemianechoic sound field, which is in tandem with the measurements performed concerning the substitution method. This way, the receivers have the same distance from the source in case it is located at the centre of the sphere or hemisphere. Apart from the surface shape, the positions of the microphones also play an important part in the sound power determination. The surface must be representatively sampled, otherwise errors may occur, such as in the case of highly directional sound sources or irregularly shaped sources. Figure 2.4 shows two different measurement surface configurations that were compared for the receivers grid determination. The first configuration consists of a coaxial circular grid of receivers. In the second, the receivers are positioned along a helicoid path. The spherical coordinates for the spiral path positioning can be calculated using Eq.(2.1) for the following elevation and azimuthal angles:

$$\begin{array}{ll} \text{Spherical helix} & \text{Hemispherical helix} \\ 0 \le \varphi_{\mathrm{az}} \le 2\pi M & 0 \le \varphi_{\mathrm{az}} \le \pi M \\ \theta_{\mathrm{el}} = \cos^{-1}\left(\dfrac{\varphi_{\mathrm{az}}}{\pi M} - 1\right) & \theta_{\mathrm{el}} = \cos^{-1}\left(\dfrac{\varphi_{\mathrm{az}}}{\pi M}\right) \end{array} \tag{2.15}$$

where M is the number of helices.

The calculation of the surface and time averaged sound pressure and sound intensity values requires attention in terms of the surface weighting, so as the partial surface to be properly sampled by each microphone [Nob99].

For a coaxial configuration of n_{rec} receivers, the surface and time averaged sound pressure level is given by:

$$\overline{L_p} = 10\lg\left[\frac{1}{S_{\mathrm{tot}}}\sum_{i=1}^{n_{\mathrm{rec}}}\frac{\frac{1}{2}\mathrm{Re}\left\{\left(p_{\mathrm{dir}}+p_{\mathrm{refl}}\right)\cdot\left(p_{\mathrm{dir}}+p_{\mathrm{refl}}\right)^{*}dS_i\right\}}{p_{\mathrm{ref}}^2}\right]\mathrm{dB} \tag{2.16}$$

where Re denotes the real part of a complex quantity, * denotes complex conjugate, S_{tot} is the total surface area, $dS_i = r^2\sin\theta_{\mathrm{el}}\, d\theta_{\mathrm{el}}\, d\varphi_{\mathrm{az}}$ is the area element covered by the i-th receiver and $p_{\mathrm{ref}} = 20\cdot 10^{-6}$ Pa.

The surface and time averaged sound intensity level is given by:

$$\overline{L_I} = 10\lg\left[\frac{1}{S_{\mathrm{tot}}}\sum_{i=1}^{n_{\mathrm{rec}}}\frac{\frac{1}{2}\mathrm{Re}\left\{\left(p_{\mathrm{dir}}+p_{\mathrm{refl}}\right)\cdot\left(u_{\mathrm{r,\,dir}}+u_{\mathrm{r,\,refl}}\right)^{*}dS_i\right\}}{I_{\mathrm{ref}}}\right]\mathrm{dB} \tag{2.17}$$

where $I_{\mathrm{ref}} = 10^{-12}\ \mathrm{W/m^2}$.

Correspondingly, the surface and time averaged sound pressure level for a helicoid configuration of n_{rec} receivers is given by:

$$\overline{L_p} = 10\lg\left[\sum_{i=1}^{n_{\mathrm{rec}}}\frac{\frac{1}{2}\mathrm{Re}\left\{\left(p_{\mathrm{dir}}+p_{\mathrm{refl}}\right)\cdot\left(p_{\mathrm{dir}}+p_{\mathrm{refl}}\right)^{*}\right\}\Big/n_{\mathrm{rec}}}{p_{\mathrm{ref}}^2}\right]\mathrm{dB} \tag{2.18}$$

The surface and time averaged sound intensity level is given by:

$$\overline{L_I} = 10\lg\left[\sum_{i=1}^{n_{\mathrm{rec}}}\frac{\frac{1}{2}\mathrm{Re}\left\{\left(p_{\mathrm{dir}}+p_{\mathrm{refl}}\right)\cdot\left(u_{\mathrm{r,\,dir}}+u_{\mathrm{r,\,refl}}\right)^{*}\right\}\Big/n_{\mathrm{rec}}}{I_{\mathrm{ref}}}\right]\mathrm{dB} \tag{2.19}$$

Eqs.(2.18) and (2.19) require only the division by the number of receivers due to the cosine weighting of elevation angles as shown in Eq.(2.15).

Figure 2.5 shows the surface and time averaged sound pressure for both coaxial and spiral configurations as shown in Figure 2.4 for various numbers of receivers. A reflecting plane with $\xi = 5 - 11j$ after Suh [Suh99] was considered on the reflections of the hemisphere base and the mirror source model was used. The sound pressure calculations were performed using Eq.(2.12) and Eq.(2.14). The sound intensity was calculated using Eq.(2.8). The parameters for the calculations were: $z_0 = 0.1\,\mathrm{m}$, $r = 2\,\mathrm{m}$ and $\Delta r = 2\,\mathrm{cm}$. The effective complex source strength was frequency independent $q = 10^{-5}\ \mathrm{m^3/s}$.

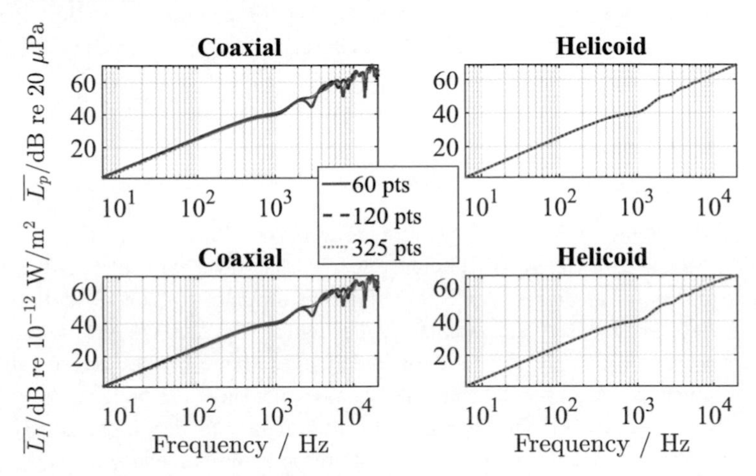

Figure 2.5:Time and surface averaged sound pressure (top) and sound intensity levels (bottom) for various numbers of receivers. Left column: coaxial receiver positioning. Right column: helicoid receiver positioning.

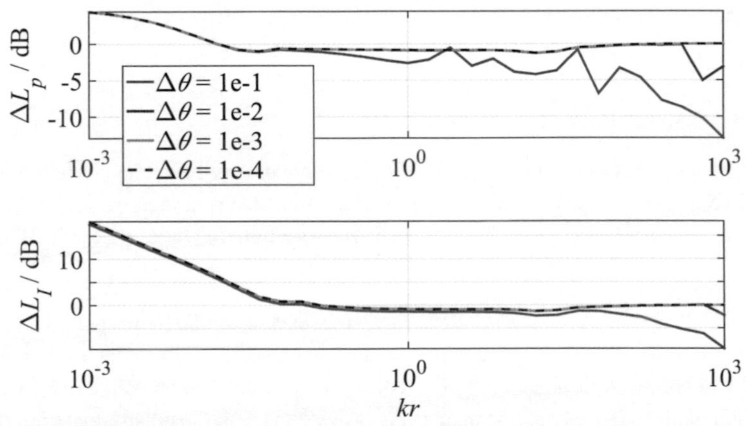

Figure 2.6:Surface and time averaged level difference for sound pressure (top) and sound intensity (bottom) between the plane and the spherical wave approach for various angle steps.

In Figure 2.5 the number of receivers influences the averaged sound pressure and sound intensity level in the coaxial configuration, since a number of microphones are placed at the same height above the reflecting plane as seen in Figure 2.4. This is not the case for the helicoid configuration. Figure 2.5 also shows that in case of helicoid configuration a smaller number of receivers is required for the same result. Based on the above, the helicoid configuration has been chosen for the further calculations of this chapter.

2.3 Angle resolution

The estimation of the surface and time averaged levels becomes more tedious in the case of the Sommerfeld integral calculation, due to the additional computational cost, mostly related to the angle step $\Delta\theta$. For the analysis provided by Suh [Suh99] and Aretz [MAr14] the proposed value was 10^{-5}. A helicoid configuration of 189 receivers above the same reflecting plane as for the calculations of Figure 2.5 was used and the averaged sound pressure and intensity level was calculated for both plane and spherical wave approach. For the latter, the following angle steps were used: $\Delta\theta = [10^{-1}\ 10^{-2}\ 10^{-3}\ 10^{-4}]$. Figure 2.6 shows the level difference between plane and spherical wave approach as described by:

$$\Delta L_{p/I} = \overline{L_{p/I\text{, plane}}} - \overline{L_{p/I\text{, spherical}}} \tag{2.20}$$

As it can be seen, the level difference remains the same for angle steps starting from 10^{-3}. This was the value used for the calculations reported in this chapter.

2.4 Sound intensity calculation

Sound intensity may be calculated either by Eq.(2.7) or Eq.(2.8). The configuration described in Figure 2.4 was used and the surface and time averaged sound intensity level over the measurement surface was estimated for both the pressure gradient and the approximation method.

The calculation of the pressure gradient was performed in Mathematica (ver. 11.1.1). Figure 2.7 shows the effect of the spacer for the pressure gradient approximation above 10 kHz for both plane and spherical wave approach. For the spherical wave approach, both sound intensity calculations provide the same results. This is not the case for the plane wave approach where the sound intensity level differs between the two intensity calculation methods. The results between plane and spherical wave approach also differ, as expected. Additionally, the computational time using the pressure derivative decreases significantly compared to the time using the gradient approximation. The spherical wave approach is more time consuming because of the Bessel function estimation. For the sound intensity calculation based on the pressure derivation, one Bessel function must be evaluated, while for the gradient approximation two Bessel functions are needed. For the above reasons, Eq.(2.7) was further used for the calculation of the sound intensity.

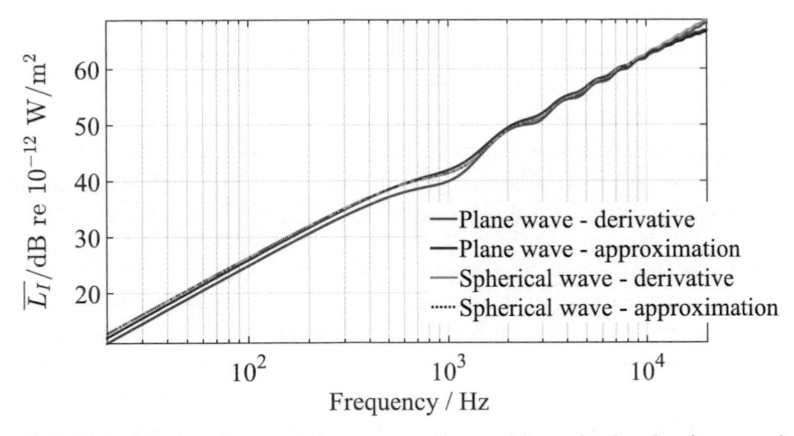

Figure 2.7: Calculated surface and time averaged sound intensity level using sound pressure derivation and approximation for plane and spherical wave.

2.5 Generalization of the results

For generalization purposes, it would be advisable to relate any frequency of interest (20 Hz – 20 kHz) to a range of radii through the Helmholtz number ($H = kr$). Similarly, any source displacement from the axis origin may be expressed as a ratio of the measurement radius

$$L_z = \frac{z_0}{r} \tag{2.21}$$

The line segments that define the translation of the source, have been taken into consideration as coordinates e.g. z_0 for translation above the x-y plane and $-z_0$ for translation below the x-y plane. This way a single equation may be used for a specific source translation. Thus, Eq.(2.12) becomes

$$p_{\text{dir}} = j\rho c k^2 q \frac{e^{-jH\sqrt{L_z^2+1-2L_z \sin\theta_{\text{el}} \cos\varphi_{\text{az}}}}}{4\pi H\sqrt{L_z^2+1-2L_z \sin\theta_{\text{el}} \cos\varphi_{\text{az}}}} \tag{2.22}$$

2.6 Translation of the substituted source

The aim of the substitution method is to determine the unknown sound power level of a source by comparing it to the known sound power of another source. The sound field quantity of sound pressure for both sources must be measured. Part of this study is the implementation of the substitution method using the sound intensity as well, which is also a field quantity. By definition, both sound pressure and sound intensity values depend on the sound environment. This dependency along with the aim to determine the sound power of the unknown source that would be emitted in free field conditions, makes the theoretical investigation of the substitution method imperative.

The theoretical calculations focused on three main aspects: i) the substitution of sources of same or different order, ii) the position of the substituted (of unknown sound power) in relation to the position of the known sound power source and iii) the existence of an impedance plane. Two sources were chosen: a monopole and a lateral dipole. All sources had the same effective complex source strength (q = 10^{-5} m^3/s). For the determination of the sound pressure and sound intensity, the dipole was implemented as the combination of two monopoles of opposite sign. Figure 2.8 shows the sources considered.

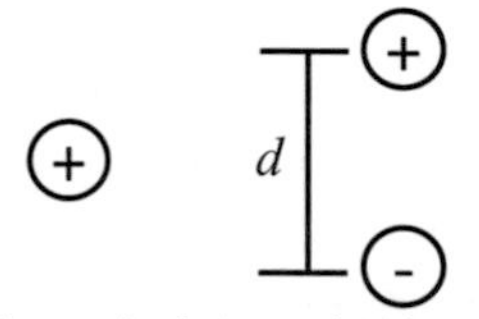

Figure 2.8: Sources for the theoretical investigation of the substitution method. Monopole (left) and lateral dipole (right).

The substitution calculations may be divided in two main categories: a) the substituted source to be of the same order as the known (monopole-monopole, dipole-dipole) and b) the substituted source to be of different order to the known (monopole-dipole). Table 2.1 summarizes the theoretical calculations configurations.

Table 2.1: Configurations for the theoretical application of the substitution method.

<table>
<tr><th>Known source</th><th>Unknown source</th><th>Sound field (measurement surface)</th><th>Translation of the unknown source</th></tr>
<tr><td>Monopole</td><td>Monopole</td><td rowspan="3">• Free field (sphere)
• Hemifree / highly reflecting floor (hemisphere). Plane and spherical wave approach for the reflected sound.
• Hemifree / highly absorbing side wall (hemisphere). Plane and spherical wave approach for the reflected sound.</td><td rowspan="3">Vertical and horizontal</td></tr>
<tr><td>Dipole</td><td>Dipole</td></tr>
<tr><td>Monopole</td><td>Dipole</td></tr>
</table>

Figure 2.9 visualizes the substitution method geometry configurations as mentioned in Table 2.1. Table A.1 in Appendix A contains the spherical coordinates of the original and mirror sources for the configurations depicted in Figure 2.9. Figure 2.10 and Figure 2.11 provide a geometry explanation of the distances from Table A.1. Following the generalization approach described in Eq.(2.21), similar replacements are:

$$L_x = \frac{x_0}{r} \tag{2.23}$$

$$Dx = \frac{\Delta x}{r} \tag{2.24}$$

and

$$D = \frac{d}{r} \tag{2.25}$$

For the derivation of the sound pressure and intensity equations, apart from the spherical coordinates of the source position, the specular reflection angle is also of great importance. Table A.2 contains the reflection angle for the geometries shown in Figure 2.9. For the calculation of r_S and $r_{S'}$ the coordinates of Table A.1 may be applied in Eq.(2.9). The angles of Table A.2 may be used for the calculation of the reflection factor of Eq.(2.4). This way the sound pressure equations for the original and the mirror source may be derived. For the reflected sound pressure according to the spherical wave approach, the values of Table A.2 may be used in the following way. The nominator of each arctan function corresponds to the parenthesis factor of the Bessel function in Eq.(2.13). Accordingly, the denominator corresponds to the parenthesis factor of the exponential function. For both plane and spherical wave approach, the particle velocity for the sound intensity calculation may be made using Eq.(2.7). For the derivation of Eq.(2.13) the following equation may be used:

$$\left[\int f(x)dx\right]' = \int f'(x)dx \tag{2.26}$$

2.7 Number of receivers and frequency resolution

Another important parameter of the theoretical calculations was the number of receivers that constitute the measurement surface over which the averaged sound pressure and sound intensity must be determined. In room acoustics modelling, a single pair of source-receiver is usually used for the calculation of the related acoustic parameters. This is not the case in the present study. Boucher [MAB16] used various half-space solutions in order to investigate interference effects. Among solutions, the Sommerfeld integral was initially included as a potential candidate. For calculations of many receiving points this solution was excluded, because of the high computational time. Ultimately, the Sommerfeld integral solution was rejected as being the most time consuming.

For the present calculations, apart from the number of receivers, the frequency resolution was also of great importance, since the difference between the plane and the spherical wave approach is higher at low frequencies. On the same frequency range, the substitution method aims to widen the applicability of the sound power determination. Thus, a finer than the one-third octave band resolution was required. The different positions of the translated source also increase the computational time. After experimental trial, the Helmholtz number vector was determined by 310 logarithmically spaced frequency lines ranging from 10^{-3} to 10^{3}. Comparison between results of the same geometry but of different frequency resolution, revealed differences in high frequencies, with the finer resolution spectra to be smoother than those with coarse resolution. The frequency resolution was decided not to be increased because it is an order of magnitude larger than one-third octave band analysis. More frequency lines would demand more computational time, which should be as limited as possible in order to make the theoretical model easy to use in case of future applicability to realistic measurements. Apparently, the same resolution applies globally to all calculations.

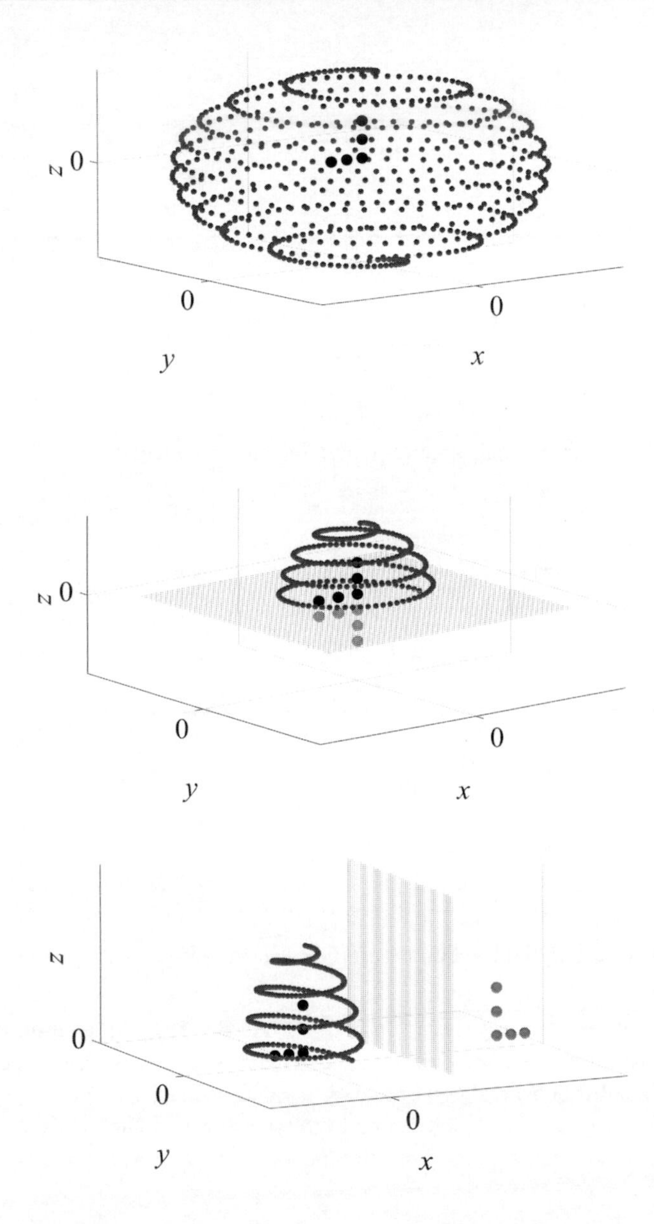

Figure 2.9: Geometrical configurations for the substitution method. Top: free field conditions. Middle: Reflecting floor. Bottom: Absorbing side wall. Vertical and horizontal translation.

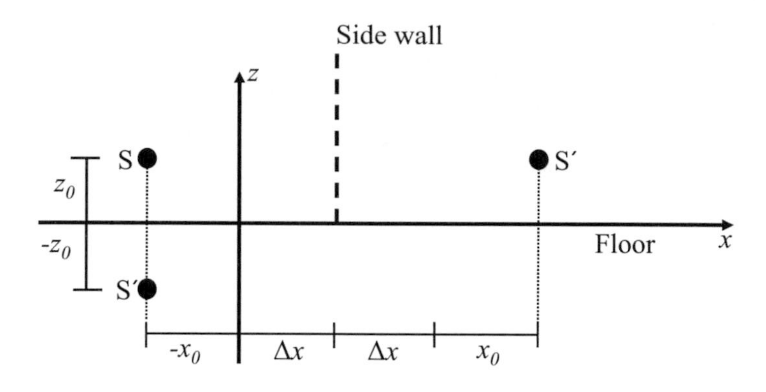

Figure 2.10: Geometry explanation for monopole mirror source.

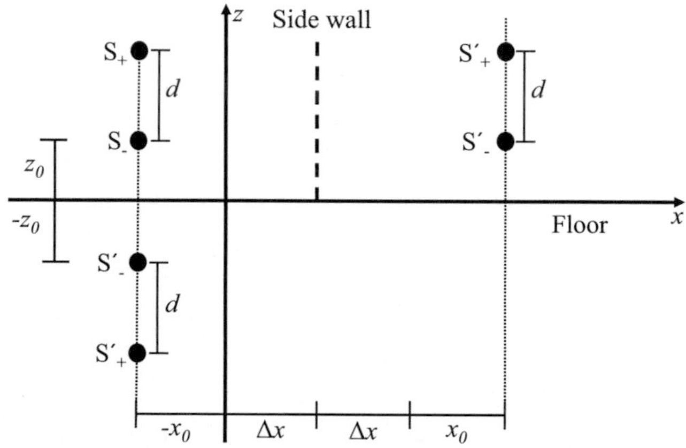

Figure 2.11: Geometry explanation for dipole mirror source.

The number of receivers was set to 126 or 252 using Eq.(2.15) for hemispherical or spherical measurement surface respectively. More specifically, for hemisphere and sphere, the number of helices was set to $M = 4$ and the azimuthal angle step was 0.1. These settings led to computational time in the range of 50 mins for the case of a vertically translated dipole over a reflecting floor (see Figure 2.9) for five translation positions.

2.8 Free field substitution

The substitution method was initially applied in free field conditions with the sources positioned within a spherical measurement surface. The known source was at the centre of the sphere whereas the unknown source was translated both vertically and horizontally as shown in the two top graphs of Figure 2.9. The free field sound pressure of the monopole is given by [Fah98]:

$$p_{\text{mon, free field}} = j\rho c k^2 q \frac{e^{-jH}}{4\pi H} \tag{2.27}$$

The radial component of the particle velocity using Eq.(2.7) is:

$$u_{\text{r, mon, free field}} = k^2 q (1 + jH) \frac{e^{-jH}}{4\pi H} \tag{2.28}$$

The free field sound pressure enables the surface and time averaged sound pressure level to be calculated following Eq.(2.18), as:

$$\overline{L_{p\text{, mon, free field}}} = 10\lg\left[\sum_{i=1}^{n_{\text{rec}}} \frac{\frac{1}{2}\text{Re}\left\{p_{\text{mon, free field}} \cdot p^*_{\text{mon, free field}}\right\} / n_{\text{rec}}}{p^2_{\text{ref}}}\right] \text{dB} \tag{2.29}$$

Similarly, the surface and time averaged level of the sound intensity normal component is:

$$\overline{L_{I\text{, mon, free field}}} = 10\lg\left[\sum_{i=1}^{n_{\text{rec}}} \frac{\frac{1}{2}\text{Re}\left\{p_{\text{mon, free field}} \cdot u^*_{\text{r, mon, free field}}\right\} / n_{\text{rec}}}{I_{\text{ref}}}\right] \text{dB} \tag{2.30}$$

In order to relate the sound power level to the generalization of the mathematical expression, the free field sound power of the monopole was calculated as:

$$L_{W\text{, mon, free field}} = 10\lg\left[\sum_{i=1}^{n_{\text{rec}}} \frac{\frac{1}{2}\text{Re}\left\{p_{\text{mon, free field}} \cdot u^*_{\text{r, mon, free field}}\right\} / n_{\text{rec}}}{P_{\text{ref}}} \frac{4\pi H^2}{k^2}\right] \text{dB} \tag{2.31}$$

The same procedure was followed for the dipole free field analysis. The free field sound pressure of the dipole is:

$$p_{\text{dip, free field}} = \frac{j\rho c k^2}{2\pi H} q \left(\frac{e^{-\frac{jH}{2}\sqrt{4+D^2-4D\cos\theta_{\text{el}}}}}{\sqrt{4+D^2-4D\cos\theta_{\text{el}}}} - \frac{e^{-\frac{jH}{2}\sqrt{4+D^2+4D\cos\theta_{\text{el}}}}}{\sqrt{4+D^2+4D\cos\theta_{\text{el}}}} \right) \tag{2.32}$$

The particle velocity was calculated for each dipole pole as the contribution of each source to the overall sound intensity [Fah95]:

$$I = \left(p_{S_+} + p_{S_-}\right)u^*_{S_+} + \left(p_{S_+} + p_{S_-}\right)u^*_{S_-} \tag{2.33}$$

The radial component of the particle velocity is:

$$\begin{aligned} &u_{\mathrm{r,\,dip,\,free\,field,\,S_+}} = \\ &k^2 q e^{-\frac{jH}{2}\sqrt{4+D^2-4D\cos\theta_{\mathrm{el}}}} \frac{\left(2 - D\cos\theta_{\mathrm{el}}\right)\left(2 + jH\sqrt{4+D^2-4D\cos\theta_{\mathrm{el}}}\right)}{2\pi H^2\left(4+D^2-4D\cos\theta_{\mathrm{el}}\right)^{3/2}} \\ &u_{\mathrm{r,\,dip,\,free\,field,\,S_-}} = \\ &k^2 q e^{-\frac{jH}{2}\sqrt{4+D^2+4D\cos\theta_{\mathrm{el}}}} \frac{\left(2 + D\cos\theta_{\mathrm{el}}\right)\left(-2 - jH\sqrt{4+D^2+4D\cos\theta_{\mathrm{el}}}\right)}{2\pi H^2\left(4+D^2+4D\cos\theta_{\mathrm{el}}\right)^{3/2}} \end{aligned} \tag{2.34}$$

Using Eq. (2.33) the averaged levels are:

$$\overline{L_{p,\,\mathrm{dip,\,free\,field}}} = 10\lg\left[\sum_{i=1}^{n_{\mathrm{rec}}} \frac{\frac{1}{2}\mathrm{Re}\left\{p_{\mathrm{dip,\,free\,field}} \cdot p^*_{\mathrm{dip,\,free\,field}}\right\} / n_{\mathrm{rec}}}{p^2_{\mathrm{ref}}}\right]\mathrm{dB} \tag{2.35}$$

and

$$\overline{L_{I,\,\mathrm{dip,\,free\,field}}} = 10\lg\frac{\left(\sum_{i=1}^{n_{\mathrm{rec}}}\left[\frac{\frac{1}{2}\mathrm{Re}\left\{p_{\mathrm{dip,\,free\,field}} \cdot u^*_{\mathrm{r,\,dip,\,free\,field,\,S_+}}\right\}}{n_{\mathrm{rec}}}\right] + \sum_{i=1}^{n_{\mathrm{rec}}}\left[\frac{\frac{1}{2}\mathrm{Re}\left\{p_{\mathrm{dip,\,free\,field}} \cdot u^*_{\mathrm{r,\,dip,\,free\,field,\,S_-}}\right\}}{n_{\mathrm{rec}}}\right]\right)}{I_{\mathrm{ref}}}\mathrm{dB} \tag{2.36}$$

The dipole free field sound power level is given by:

The dipole free field sound power level is given by:

$$\overline{L_{W\text{, dip, free field}}} = 10\lg\frac{\left(\sum_{i=1}^{n_{\text{rec}}}\left[\frac{\frac{1}{2}\text{Re}\left\{p_{\text{dip, free field}}\cdot u^{*}_{\text{r,dip,free field, S}_{+}}\right\}}{n_{\text{rec}}}\right]+\sum_{i=1}^{n_{\text{rec}}}\left[\frac{\frac{1}{2}\text{Re}\left\{p_{\text{dip, free field}}\cdot u^{*}_{\text{r, dip, free field, S}_{-}}\right\}}{n_{\text{rec}}}\right]\right)\times\frac{4\pi H^2}{k^2}}{P_{\text{ref}}}\text{dB} \quad (2.37)$$

The substitution method was implemented for all cases described in Table 2.1 for the following translation increments: L_x = [-0.1 -0.2 -0.3 -0.4 -0.5] and L_z = [0.1 0.2 0.3 0.4 0.5] (see Figure 2.9). The sound power of the unknown source after applying the substitution method was compared to the free field sound power according to:

$$\Delta L_W = L_{W\text{, sub}} - L_{W\text{, free field}} \quad (2.38)$$

Figure 2.12 shows the sound power level differences for all configurations considered. If both unknown and known source are monopoles, the substitution method provides deviations up to 0.47 dB when sound pressure is used. The deviations become negligible if sound intensity is used. The use of sound pressure becomes critical in case at least a dipole is included. The deviations can be up to 3 dB for vertical translation of the unknown source and 2.5 dB for horizontal translation. Sound intensity is also in this case insensitive to the translation of the unknown source. The substitution method yields large deviations in the case where the sources are different (known: monopole, unknown: dipole) when sound pressure is used. This is not the case when sound intensity is used. The deviations remain up to 0.2 dB for horizontal translation. In overall, Figure 2.12 justifies the use of sound intensity for the substitution method, since it provides results close to the free field values for all configurations for which the calculations were performed.

2.9 Substitution including an impedance plane

The substitution method was also applied for the case of an impedance plane. Figure 2.9 shows the translation of the unknown source for the case of a reflecting floor and a reflecting side wall. Compared to free field calculations, the reflection factor must be included in the calculations. This includes the specific acoustic impedance ξ, which for a locally reacting boundary is independent of the incidence angle [Kut79].

On the other hand, the reflection factor depends on the cosine of the reflection angle θ as described by Eq.(2.4). According to Kuttruff [Kut79] the absorption coefficient is related to ξ and the reflection angle for the case of oblique incidence (see Figure 2.3) through:

$$\alpha(\theta_0)=\frac{4\,\mathrm{Re}\{\xi\}\cos\theta_0}{\left(|\xi|\cos\theta_0\right)^2+2\,\mathrm{Re}\{\xi\}\cos\theta_0+1} \tag{2.39}$$

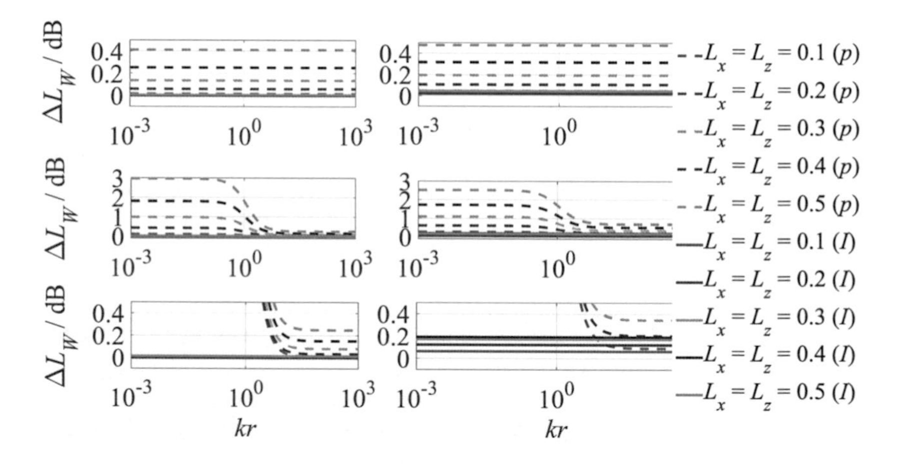

Figure 2.12: Sound power level difference between level after substitution method and free field level using both sound pressure (*p*) and sound intensity (*I*). Left column: vertical translation of the unknown source (L_z). Right column: horizontal translation of the unknown source (L_x). Top row: known source: monopole, unknown source: monopole. Middle row: known source: dipole, unknown source: dipole. Bottom row: known source: monopole, unknown source: dipole.

The product of the impedance and cosine of the reflection angle strongly affects the field quantities that describe the reflected sound. The selection of an impedance value to be used for all intended calculations, e.g. sound pressure, sound intensity, monopole, dipole etc., becomes non-detrimental. In bibliography ([MAr14], [Bie03]), three different impedance values were used to cover low, moderate and high absorption values, namely 5-11j (α = 0.21), 1-2.83j (α = 0.41) and 0.59+0.57j (α = 0.71). The calculation of the surface and time averaged sound pressure level using the previously described models, lead to results that come in accordance with theory, for some kr, especially in the near field, sound intensity levels are larger than the corresponding sound pressure levels for both plane and spherical wave approach, which comes in contradiction to the sound intensity theory. In realistic sound intensity measurements, the case of larger sound intensity levels than sound pressure levels is rejected as erroneous. In the present study, since the inclusion of the sound intensity for the determination of the sound power using the substitution method is a novelty, which among other means lack of bibliography, the substitution calculations were further performed, since the full development of the models for the impedance plane lies outside the scope of this study, but may be an interesting topic for future work.

Before presenting the results of the substitution method, a list of observations for the not fully correct sound intensity results is given. In Eq.(2.14) the specific acoustic impedance is included once in the nominator and once in the denominator. In case of the derivation for the particle velocity calculation, the impedance becomes apparent to more terms, making the interpretation of its influence complicated. It has been observed that for a given value of the specific acoustic impedance, the model gives correct sound intensity levels (lower than the sound pressure levels) for the entire frequency range in case of a dipole but not for the monopole. This may lead to the assumption that the distance of the dipole poles and the distance source-reflecting plane is also influential. Larger sound intensity levels than sound pressure levels were also observed for the case where the particle velocity was calculated based on the pressure gradient approximation.

For the case of the translated dipole against a side wall, the sound pressure level does not vary while varying the impedance of the plane. This is attributed to the low side directivity of the dipole. On the contrary, this is not the case for the sound intensity calculation. For the case of the reflecting floor, the specular reflection angle spans from 0.4° to 84°. For the absorbing side wall, the angle varies from 0.4° to 21°.

Based on the above observations, the specific acoustic impedance was chosen after trial and error. The major criterion for the selection of the impedance was to yield more correct sound intensity values for monopoles and dipoles. For the case of the highly reflecting floor, the impedance was set to $15+50j$ and for the absorbing side wall $0.6+1.4j$. The absorption coefficient for both cases is shown in Figure 2.13 as sets for all configurations. The coefficient is shown as a function of the measurement surface point, because the reflection angle slightly changes by changing the source position (vertical and horizontal translation). The highest possible absorption coefficient (0.54) was also related to the impedance value and the related sound intensity implications.

The substitution method was performed in the way described in chapter 2.8 and the results were compared to the free field sound power using Eq.(2.38). Additionally, the comparison to the free field sound power was performed for the sound power levels determined by directly using the surface and time averaged sound pressure and intensity levels using the following equations [Ver06]:

$$L_W = \overline{L_p} + 10\lg\left(\frac{2\pi H^2}{k^2 S_0}\right)\text{dB} - 10\lg\left(\frac{\rho c}{400}\frac{\text{m}^3}{\text{kg}\cdot\text{s}}\right)\text{dB} \tag{2.40}$$

$$L_W = \overline{L_I} + 10\lg\left(\frac{2\pi H^2}{k^2 S_0}\right)\text{dB} \tag{2.41}$$

where $S_0 = 1\text{m}^2$.

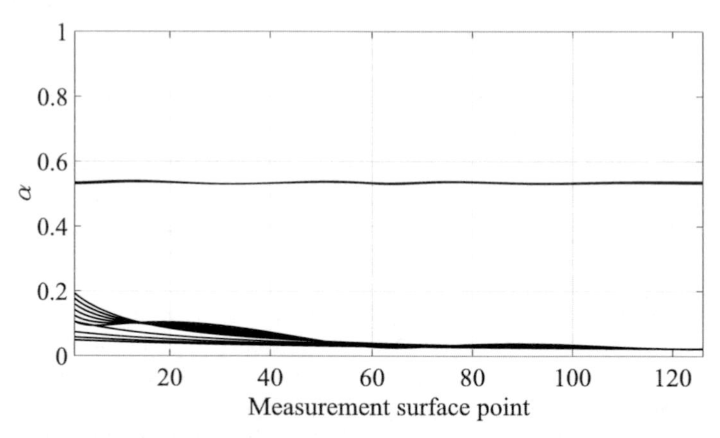

Figure 2.13: Absorption coefficient for calculations including impedance planes. Higher value lines: absorbing side wall configurations. Lower value lines: reflecting floor configurations.

Figure 2.14 shows the surface and time averaged sound pressure and intensity levels for a horizontally translated monopole ($L_x = 0.1$) in case of a highly reflecting plane.

As it may be seen, in the near field ($kr < 1$) the levels differ and the sound intensity levels become higher than the corresponding sound pressure levels. On the other hand, in the far field ($kr > 1$) the levels of each quantity converge, while the sound intensity levels are lower than the sound pressure levels.

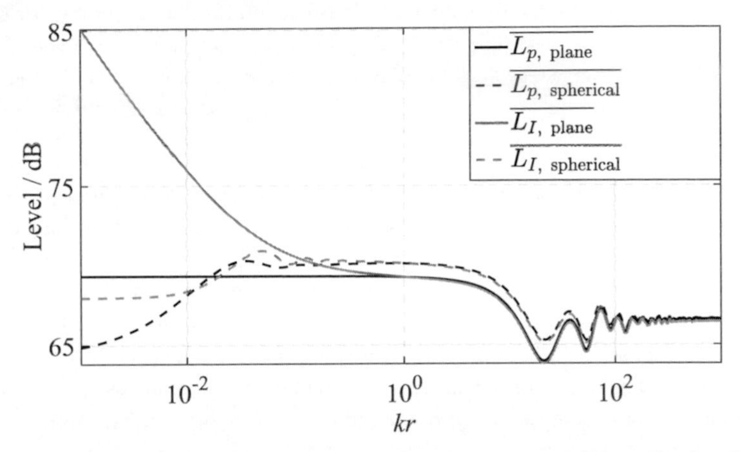

Figure 2.14: Surface and time averaged sound pressure and intensity level for a horizontally translated monopole over a reflecting plane.

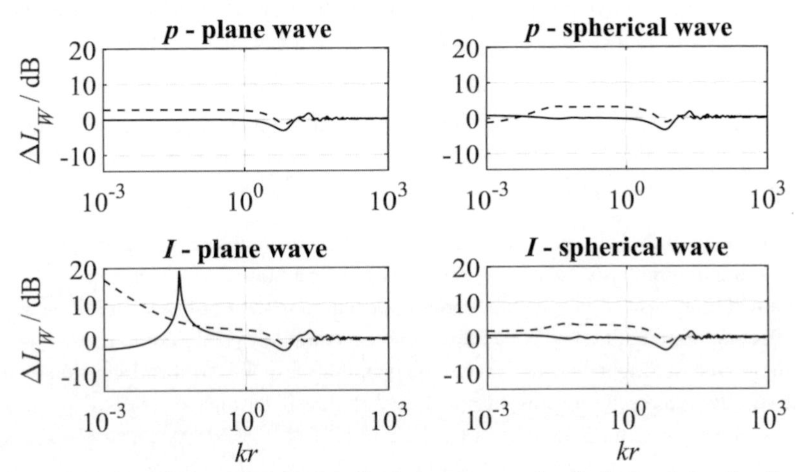

Figure 2.15: Deviation from the free field sound power level after applying the direct (dashed) and the substitution method (continuous). Known source: monopole. Unknown source: monopole. Vertical translation over reflecting floor. $L_z = 0.3$.

The deviation of the sound power levels after applying both the direct and the substitution method from the corresponding free field sound power level have been derived using Eq.(2.38). Figure 2.15 shows the deviation for all field quantities and wave approaches used for the sound power determination for the case of a vertically translated monopole over a highly reflecting floor. As indicative, the translation distance (L_x, L_z) was set to 0.3, since for all translations similar findings can be found. For the side wall configurations the normalized distance from the axis origin to the side wall was $\Delta x = 1.5$.

As it can be seen in Figure 2.15 both sound power determination methods produce similar results for large *kr* (>$10^{1.8}$), while the influence of the reflecting plane is apparent to the direct method sound power levels (~ 3 dB) for small *kr*. This exhibits that the substitution method gives the free field sound power and the direct method the sound power that is truly emitted by the source. It must also be stated that the substitution method gives the exact free field results in case both sources are placed at the same position. The large deviations for sound intensity results using the plane wave approach for very low *kr* are attributed to the distance between the original and mirror source.

In Appendix A the deviations for the cases described in Table 2.1 are shown. Similar results to those of Figure 2.15 can be seen in all figures of Appendix A that describe the substitution of a monopole by a monopole. In the case of the absorbing side wall the deviation from the free field sound power is smoother due to the inexistence of strong reflections.

The applicability of the substitution method is enhanced for the case of the substitution of a dipole by a dipole. In all cases, the near field deviation of the levels after the substitution method is much lower than the direct method.

The deviation from the free field sound power levels becomes large, especially in the near field, for both sound power determination methods for the case where the unknown source is a dipole and the known one a monopole. For the calculations with a reflecting floor, both sound power determination methods yield to similar deviations from the free field sound power level. This is not the case for the absorbing side wall. The substitution results are closer to the free field levels for high *kr*, with the range to vary with varying configuration and field quantity.

When the substitution method involves sources of the same order, the spherical wave approach gives sound power values closer to the free field ones by using both sound pressure and intensity for all *kr*. When a dipole is substituted by a monopole the difference between the plane and spherical wave approach is not as large as previously, with the substitution results using sound intensity and plane wave approach to give a wider *kr* range with small deviations from the free field sound power.

The standard deviation of the deviations from the free field sound power levels for the substitution method results reveals that the sound intensity based on the spherical wave approach gives results not so widespread as the sound pressure using the same approach. The same can be seen for the substitution of a dipole by a dipole except for the case of vertical translation over floor. The lowest spread of deviations from the free field levels concerning the substitution of a dipole by a monopole depends on the geometry and the distance from the source.

A general result for the substitution method including a reflecting plane is that the horizontal translation of the unknown source with respect to the known source produces lower deviations from the free field sound power level for a wide range of the spectrum.

2.10 Geometry with two reflecting planes

An attempt was made for the application of the substitution method in a geometry, which includes both a floor and a side wall. Compared to the previous calculations, this means that the integral evaluation range of Eq.(2.13) must be changed. The limit of the real angles is no more $\pi/2$ but it is determined by the intersection point of the two planes. This change of the angle has an effect on the Sommerfeld integral evaluation as shown in the Appendix of [Suh99], because the simplification of the identity used in Eq.(A3) is no longer applicable and more trigonometric functions (sin, cos, sinh, cosh) are introduced in the integral evaluation. These extra functions significantly affect the convergence of the integral using imaginary angles. Apparently, the results of the two planes calculations indicated that further investigation is required, which lies outside the scope of the present study, but could be the topic for future analysis.

The theoretical investigation of the substitution method opens the way for the experimental part of the study. Before providing measurement results related to the substitution method, an apparatus for the determination of the time and surface averaged sound pressure and sound intensity levels is presented. A detailed discussion is performed on features of the apparatus, which are of importance for the sound power determination. The apparatus was used for the determination of the sound power of the transfer standar under calibration conditions. The related uncertainty was also determined and decomposed to the contributing factors.

3 Calibration procedure for transfer standards

In sound power measurements the use of reference sound sources is based on ISO 6926 [ISO6926], which describes the proposed procedure as calibration. Although the wide acceptance of the procedure by the acoustical community, according to the international vocabulary of metrology [JCG12], it is a reference measurement procedure. Calibration is defined as the operation, which relates the quantity values (in our case the unit watt) with measurement uncertainties provided by measurement standards [JCG12]. In the context of this study, a primary standard (primary source) and several transfer standards (transfer sources) were used. The sound power of the latter can be referred to the sound power of the former through the dissemination process. The measurements and uncertainty calculation required, for the transfer standards calibration procedure, are described in the following paragraphs.

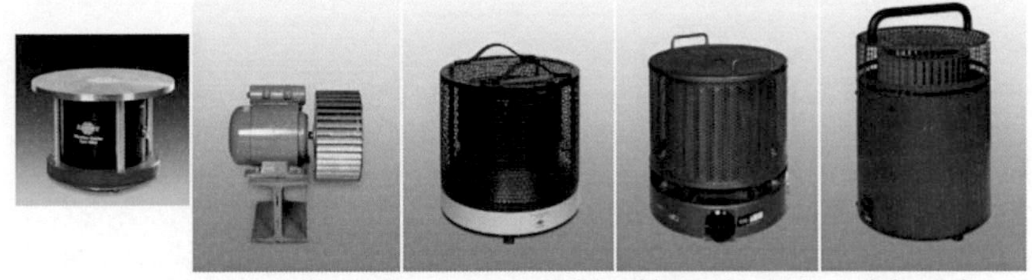

Figure 3.1: Primary and transfer sources used in this study. Left to right: primary source, AEG Type ADEB71K2, Brüel & Kjær Type 4204, EDF Type A and Norsonic Type Nor 278.

PTB (the National Metrology Institute of Germany) designed and assembled a primary source. The source consists of a vibrating piston, a shaker and a housing construction allowing the source to be embedded on the PTB's hemianechoic room floor while isolating the surrounding plate from vibrations [Kir06]. As transfer sound sources, six aerodynamic reference sound sources (RSSs) were used. This type of sources was chosen because they are widely used in sound power determination. In a RSS the sound is generated by a fan driven by an asynchronous motor ([Har59], [Fra77], [Cam91], [Suz93], [Cam08])]. Six RSSs were used of four different types, namely an AEG Type ADEB71K2 (AEG), three Brüel & Kjær Type 4204 (B&K), an EDF Type A (EDF) and a Norsonic Type Nor 278 (NOR). Apart from this type of sources, other types have also been developed ([Tak00], [van18]). Figure 3.1 shows the primary and the transfer sources used for the measurements of this study.

3.1 Scanning apparatus

Sound power is determined by field quantities (sound pressure or sound intensity) measured either at free or diffuse sound field (or at approximations of those). In case of free field, the measurements are performed over a surface. The measurements may utilize stationary microphones, scanning intensity probes (usually hand held) or a mechanical system that moves a microphone along a meridional path over the source under test [ISO3745]. The latter technique is used by many acoustical laboratories, including National Metrology Institutes. The quarter circle moving path of the microphone along with the simultaneous rotation of the source enables the sound power determination in a virtual hemisphere of varying radius. There is a main disadvantage for this kind of measurements. The microphone moves along the same field points relative to the room, which has not perfect frequency response at low frequencies.

In order to overcome these disadvantages and to gain better insight to effects, which affect the sound power determination (e.g. near field effects) a scanning apparatus was specially assembled by PTB. The scanning apparatus consists of two parts, a stainless steel semicircular arc placed in PTB's hemianechoic room and two motors positioned outside the room, which enable the moving of the arc back and forth. There are two arcs for different radii. Each arc has holes where up to 24 microphones can be mounted. Figure 3.2 shows both arcs of the scanning apparatus with the largest bearing all microphones possible. At the centre of the semicircles, the spot at which the sources under test can be positioned is visible.

The microphones are attached to one edge of 70 cm long acrylic glass rods. The manual movement of the rods along the arc holes enables the variation of the measurement radius. The same kind of rods with edge modification can be also used for the attachment of sound intensity probes. The hole positions were chosen to cover equal surface by each microphone in case of 24 microphones. The angles of the microphones positions with respect to the floor are: 17°, 29°, 38°, 44°, 51°, 57°, 63°, 68°, 73°, 78°, 83°, 88°, 92°, 97°, 102°, 107°, 112°, 117°, 123°, 129°, 136°, 142°, 151° & 163°. Figure 3.3 shows a microphone positioned on the arc using a rod.

The arc is tilted by two motors with one reel each, which control the length of two wire ropes that transmit the movement to the arc. The motors are attached to a massive concrete block so that their vibrations are isolated and not transmitted to the hemianechoic room side wall. The wire ropes length along with the arc weight may induce arc vibrations during scan. For this reason, each wire rope is not directly attached to the arc but to a rope, which in turn is attached to the arc. Between them springs are intersected. The scan parameters are set by software and the scan speed is measured by a potentiometer.

Figure 3.2: Scanning apparatus.

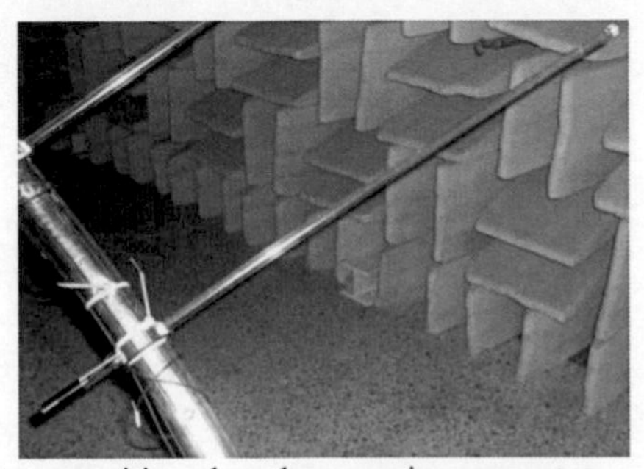

Figure 3.3: Microphone positioned on the scanning apparatus.

3.2 Measurable quantities

One scope of this study is to apply the substitution method using sound intensity as well. For this reason, the scanning apparatus has been designed that both sound pressure and sound intensity measurements can be performed.

3.2.1 Sound pressure

For the measurement described in this study 1/2" free field microphones were used (G.R.A.S. Type 40AF) with 1/4" preamplifiers (G.R.A.S. Type 26AC-S4). The selection of these preamplifiers was made for mechanical reasons, because the subsequent adaptors enabled the attachment of the microphones to the acrylic glass rods.

The microphone signals were recorded in real time while the analysis was performed either also in real time or in post processing depending on the number of channels used. An OROS multianalyzer recorder (OR38) and NVGate software (ver. 8.30) were used for the recording and analysis of the signals. The final frequency domain data were further analyzed using Matlab (vers. 2013a – 2015a).

3.2.2 Sound intensity

Apart from the sound pressure, sound intensity can be also measured using the scanning apparatus. For this study the related measurements were performed using three intensity probes and sound intensity microphone pairs (Brüel & Kjær Type 4181 and 4197). Two spacers of different length were used (9 mm & 71 mm) to cover the frequency range between 20 Hz and 10 kHz.

The same analyzer was used as for the sound pressure measurements and the radial component of the sound intensity was calculated according to:

$$I_{\mathrm{r}} = \frac{1}{\omega_{\mathrm{ang}}\,\rho\,\Delta x}\,\mathrm{Im}\left[S_{\mathrm{p_1\,p_2}}\left(\omega_{\mathrm{ang}}\right)\right] \qquad (3.1)$$

where $\mathrm{Im}\left[S_{\mathrm{p_1\,p_2}}\left(\omega_{\mathrm{ang}}\right)\right]$ is the imaginary part of the cross spectrum of the intensity probe microphone signals [Ver06].

3.3 FFT windowing effects

The scanning apparatus has been used for the measurement of both primary and transfer sources. The primary source was driven by a fixed phase multisine signal, which exactly matches the time window of the analysis. Therefore, a uniform FFT window was applied for the measurements of the primary source. Due to the broadband characteristics and the randomness of the aerodynamic reference sound source signal, a Hanning FFT window was used. The Hanning window side lobes rise the power spectrum values by 50% compared to the same values using uniform window. In level expression, the added value is 10lg(1.5) dB = 1.76 dB [Brü87].

3.4 Background noise contribution

A desired feature of the scanning apparatus is low background noise. The background noise was measured while scanning at different scan speeds (scan duration 600 s, 900s & 1200 s). The sound pressure level of three RSSs (B&K, EDF, NOR) was also measured at various radii and compared to the background noise level for both one-third octave band and FFT (6401 lines, 3.125 Hz) frequency resolution. Figure 3.4 shows the comparison of surface and time averaged sound pressure levels.

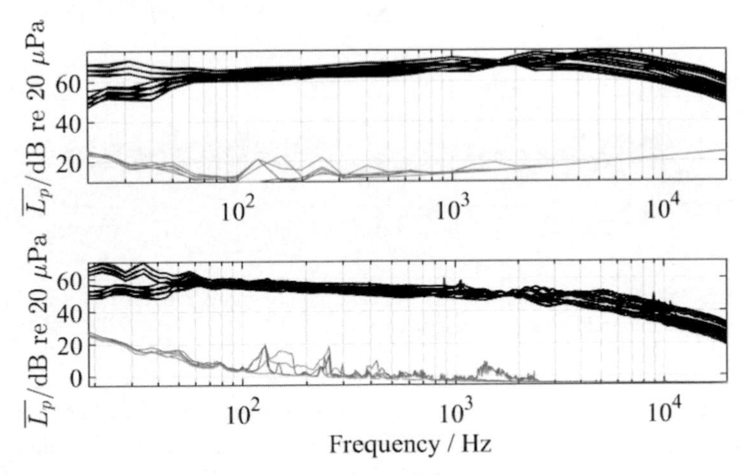

Figure 3.4: Aerodynamic reference sound sources sound pressure level (black) and background noise (grey) for one-third octave bands (top) and FFT bands (bottom).

The background noise levels are at least 20 dB lower than the sound pressure levels at which the RSSs normally operate. The effect of the increased scan speed can be observed at the region between 1 kHz and 2.5 kHz without affecting the signal to noise ratio. Apparently, the scanning apparatus background noise is unlikely to affect the RSS measurements.

3.5 Scan speed

The scan must be uniform with an adjustable duration. The latter can be achieved by the software that controls the motors. A potentiometer monitors the angular velocity of the scan. It is connected to one arc end and the rotation of the arc rotates the potentiometer as well. The slope of the potentiometer output voltage is related to the scan speed. Measurements were performed with 1200 s duration including both onwards and backwards scan. The slope of the voltage output was calculated for time intervals of 10 s. The results are presented in Figure 3.5.

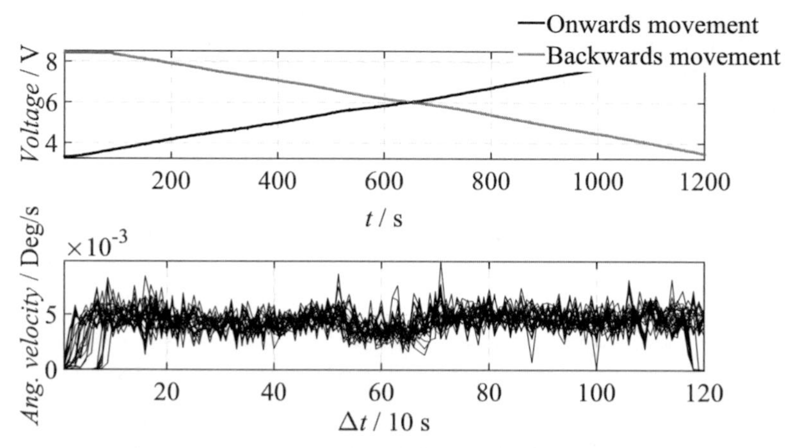

Figure 3.5: Potentiometer voltage for both scanning apparatus movements (top) and corresponding angular velocity (bottom).

As it may be seen in Figure 3.5 the scan speed has an average value of 0.005 Deg/s for most of the measurement duration. Different values may be seen at the beginning and end of the measurement and the middle of the scan. The connection of the potentiometer to the arc can explain the speed values at the beginning and end. A small plastic tube connects the potentiometer to the arc, whose flexibility prohibits the simultaneous rotation of the two. When the arc reaches its maximum position the scan speed reduces because of the reels stop and go. The measurement results repeatability can be used as proof for the sufficient stability of the scanning apparatus.

3.6 Scan repeatability

The scan speed is related to the scan repeatability. The surface and time averaged sound pressure level was measured for three different RSSs (B&K, EDF, NOR) and measurement radii. For the measurements at the same radius the standard deviation was calculated and the results are presented in Figure 3.6. The large standard deviation values at 10 kHz are attributed to the tonal characteristics of a source under investigation. It is concluded that the measurement repeatability is high due to the resulted standard deviation, which also includes the emission repeatability of the sources under investigation.

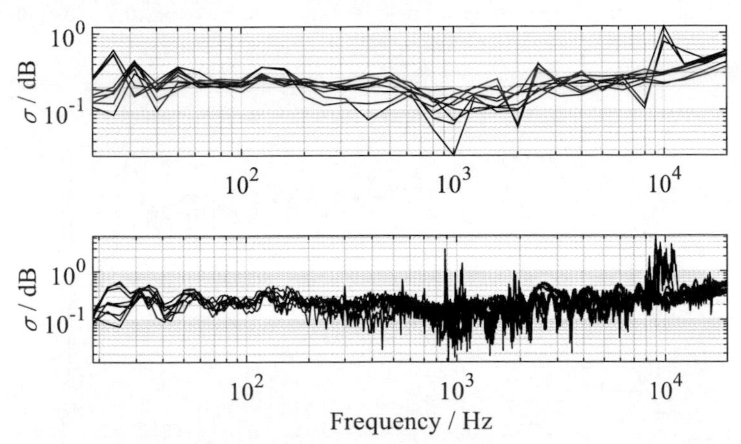

Figure 3.6: Standard deviation of the surface and time averaged sound pressure level for various radii measurements for one-third octave bands (top) and FFT bands (bottom).

3.7 Apparatus reflections

During the scan, reflections from the arc are included in the recorded signals. To investigate this, the primary source was measured using both the scanning apparatus at 2m and the meridional path mechanism at 1.91m. The measurement radii correspond to the same distance between the microphones and the equipment. The influence of reflections at high frequencies for both measurements are shown in Figure 3.7. At frequencies below 100 Hz, the influence of the room modes can be observed in both measurement mechanisms. In the frequency range between 800 Hz and 5000 Hz, it can also be observed that there are different reflections between the measuring mechanisms.

The substitution method includes the subtraction of the sound pressure levels of the two sources involved, which is expected to compensate for the arc reflections. This was investigated experimentally. Three RSSs (B&K, EDF, NOR) and the primary source were measured at three different radii (1.45 m, 1.70 m & 2.00 m). The measurement data were then referred to the same distance and the sound pressure level subtraction included in the substitution method was applied. The standard deviation of the originally measured sound pressure levels was compared to this of the level differences. The comparison is shown in Figure 3.8. The measured data is strongly affected by the arc reflections between 1250 Hz and 2000 Hz. This is compensated after the sound pressure levels subtraction for both frequency analyses. The peak at 2500 Hz is attributed to the frequency response of the primary source. It is evident that the expected reflections compensation is achievable by using the substitution method.

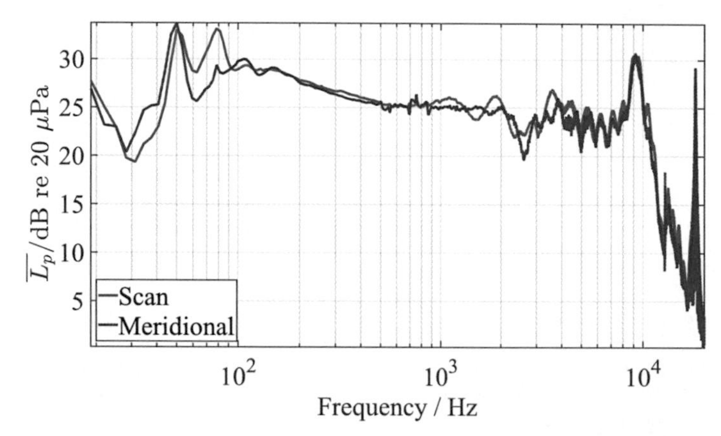

Figure 3.7: Primary source sound pressure level after different measurement techniques. Microphones at the same distance from measuring equipment.

3.8 Comparison to meridional path

The sound power determination using the scanning apparatus and the meridional path was compared in terms of level difference and measurement repeatability. Sound pressure measurements were performed for three RSSs (B&K, EDF, NOR) and the primary source at three different measurement radii with the scanning apparatus and a meridional measurement for each source. The sound power was calculated according to ISO 6926 [ISO6926]:

$$L_W = \overline{L_p} + 10\lg\left(\frac{S}{S_0}\right)\text{dB} + C_1 + C_2 + C_3 \tag{3.2}$$

where $\overline{L_p}$ is the surface and time averaged sound pressure level, S the measurement surface in m^2, $S_0 = 1\text{m}^2$, C_1 is the reference quantity correction, C_2 is the acoustic radiation impedance correction and C_3 is the correction for air absorption [ISO6926].

After the determination of the sound power level for each measurement technique, the difference between the two techniques was calculated as:

$$\Delta L_W = L_{W,\text{scan}} - L_{W,\text{meridional}} \tag{3.3}$$

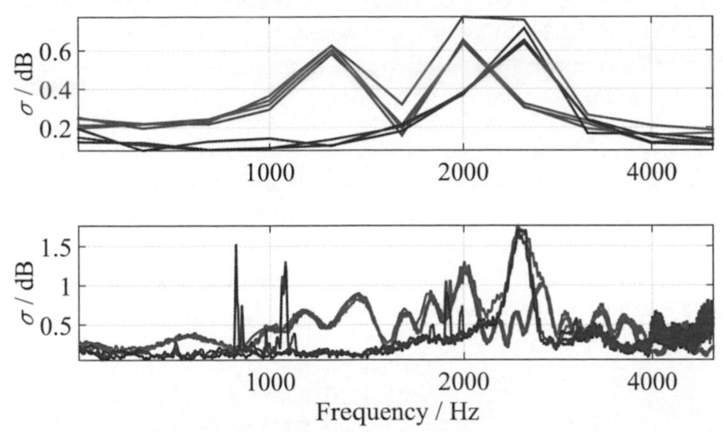

Figure 3.8: Standard deviation of the originally measured sound pressure levels (blue) and the sound pressure level difference (orange red) for one-third octave bands (top) and FFT bands (bottom).

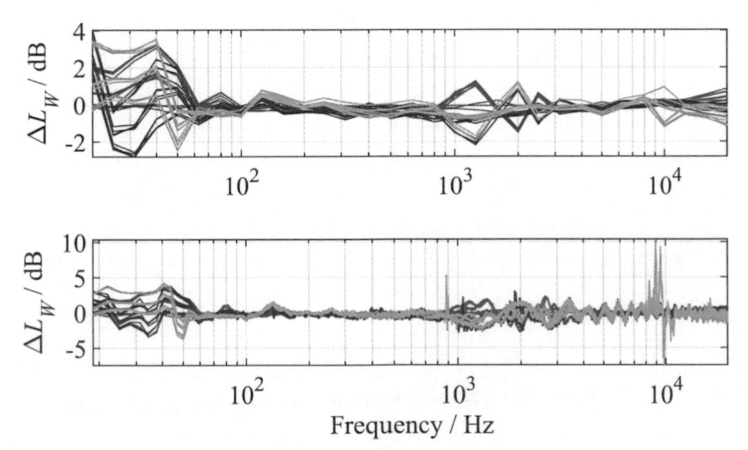

Figure 3.9: Sound power level difference between scan and meridional path measurements for one-third octave bands (top) and FFT bands (bottom). Each colour corresponds to a different reference sound source.

Figure 3.9 shows the sound power level differences for all sources. The two methods provide different sound power values. At low frequencies room modes and wind influences cause large differences. The increase at 10 kHz is due to the frequency content of a RSS. The aforementioned reflection influences are also apparent. In frequency regions where the previous phenomena are not apparent, the difference is between -2.18 dB and 1.24 dB for one-third octave band analysis and -2.76 dB and 5.21 dB for FFT measurements. The differences may also provide an evidence for the implementation of traceability of the unit watt.

The repeatability of the measurement methods was checked by comparing the standard deviation of the sound power determined by each method. PTB poses sound power levels of a RSS using the meridional path technique for more than 20 years. The standard deviation of this data was compared to the standard deviation of measurements obtained with the scanning apparatus for the same source using ISO 3745 [ISO3745]. Figure 3.10 shows the compared standard deviations.

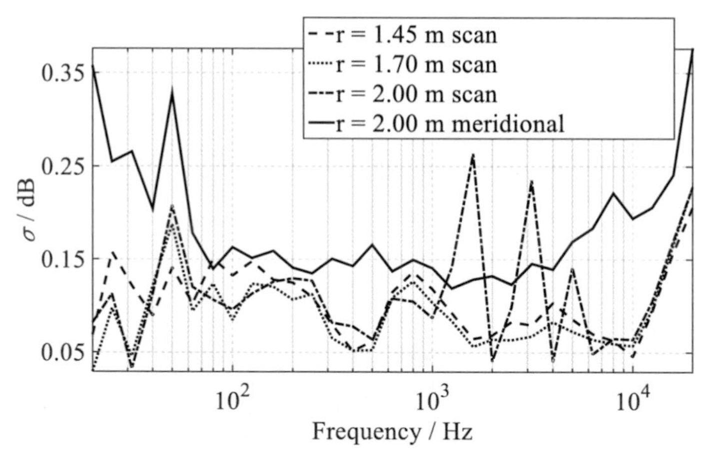

Figure 3.10: Standard deviation of sound power levels determined by the meridional path (continuous line) and the scanning method (dashed lines) for one-third octave bands.

In Figure 3.10 it can be seen that the scanning method provides better repeatability for directly determined sound power (no substitution performed), which fully supports the application of the scanning apparatus to sound power measurements. More prominent is the standard deviation decrease for the scan method at the low and high frequency end. The former is assumed to be due to the microphone paths, which may smoothen room mode effects. The latter is attributed to the source directivity, which can be better detected by the scanning apparatus. The arc reflections are to be compensated when the substitution method is applied yielding even better repeatability.

3.9 Calibration uncertainty

According to GUM [JCG08], the objective of a measurement is to determine the measurand (in this case the surface and time averaged sound pressure level), whose lack of knowledge of the exact value is described by the related uncertainty. The sound pressure measurement using the scanning apparatus for the calibration of the transfer standards must include corrections for: the filter analysis software, the changes in the sound emission by the source, background noise, the reflections from the scanning apparatus, the attenuation of the scanning apparatus in case the microphone position is above the apparatus, the microphone sensitivity, the microphone calibration, the positioning of the microphones, the sound incidence angle, FFT windowing, near field and windscreens. The calibration includes the application of the substitution method, which means that the sound pressure must be twice measured, once for the primary and once for the transfer source.

The corrected sound pressure level of the primary standard, following the approach of a previous study [Wit051], may be described by:

$$L_{p,\mathrm{PS},i} = L'_{p,\mathrm{PS},i} + C_{\mathrm{fil}} + C_{\mathrm{em}} + C_{\mathrm{noise}} + C_{\mathrm{ref}} + C_{\mathrm{att}} + C_{\mathrm{mic},i} + C_{\mathrm{cal},i} + C_{\mathrm{pos},i} + C_{\mathrm{ang},i} \qquad (3.4)$$

Similarly, the corrected sound pressure level of the transfer standard is:

$$L_{p,\mathrm{TS},i} = L'_{p,\mathrm{TS},i} + C_{\mathrm{fil}} + C_{\mathrm{em}} + C_{\mathrm{noise}} + C_{\mathrm{ref}} + C_{\mathrm{att}} + C_{\mathrm{FFT}} + C_{\mathrm{nf}} + C_{\mathrm{scr}} + C_{\mathrm{mic},i} + C_{\mathrm{cal},i} + C_{\mathrm{pos},i} + C_{\mathrm{ang},i} \qquad (3.5)$$

In the above equations C_{fil} is the correction for the filter analysis software, C_{em} the correction for the changes in the sound emission by the source, C_{noise} the background noise correction, C_{ref} the correction for the reflections from the scanning apparatus, C_{att} the correction for the attenuation of the scanning apparatus, $C_{\mathrm{mic},i}$ the correction for each microphone sensititvity, $C_{\mathrm{cal},i}$ the correction for the calibration of each microphone, $C_{\mathrm{pos},i}$ the correction for the positioning of each microphone, $C_{\mathrm{ang},i}$ the sound incidence correction for each microphone, C_{FFT} the correction for the FFT windowing, C_{nf} the correction for the near field and C_{scr} the windscreen correction.

Eqs.(3.4) and (3.5) include correction factors that either affect each microphone to the same extend or affect each microphone individually as indicated by the *i*-index. Due to the same effect of some corrections to both primary and transfer sources, these corrections will cancel out by applying the substitution method. In the following paragraphs the correction factors are explained and the contribution to the combined uncertainty are further analyzed.

3.9.1 Corrections to cancel out during substitution

The correction factor C_{fil} describes the influences of the filters applied for the frequency analysis of the data. The filter specifications may be found in the related standard [IEC14] and an uncertainty estimation in a previous study [Wit052]. The use of the same measuring equipment during all sound pressure measurements makes this correction factor vanishing during the substitution method.

The sound emission for every source under test depends on several factors (e.g. the input voltage) and the related correction is described by the factor C_{em}. The substitution method is used to describe the sound generated by the source without intervening in the sound generation process, which is in the interest of the manufacturer. The plug and play use of the sources excludes the correction for the changes in the emission from the substitution method.

The presence of the arc is related to reflections which are apparent in the sound pressure measurements as has already been examined in paragraph 3.7. It has also been presented that the substitution method results in the cancellation of the reflection influences in case the measurements are performed at the same measurement radius. For this reason, this correction is not influential for the substitution method. The same applies for the attenuation from the arc.

The free field sensitivity of the microphone is a function of frequency and the correction $C_{\mathrm{mic},i}$ may be acquired by the frequency response graph provided by the microphone manufacturer. The use of the same microphones for the measurement of the primary and the transfer standard under the assumption of linearity, leaves the correction out of the substitution application.

During the microphone calibration process before the measurements, the differences in the calibration values set by meteorological conditions are introduced by the $C_{\mathrm{cal},i}$ factor. Since the calibration is performed before both primary and transfer standard measurements, the correction factor is not taken into account for the substitution method.

In an ideal free field it is straightforward to calculate the sound pressure level at a distance r from the source. In a hemianechoic room there are room modes, which influence the measurements and a correction must be applied through the factor $C_{\mathrm{pos},i}$. The calculation of the factor for the case of a point source is presented in [Wit051]. The fixed position of the arc in the hemianechoic room leads to the same remaining wall reflections during the measurements for the substitution method and thus, the rejection of the correction as influential to the uncertainty analysis. The factor would also be important in case the frequency content of the sources was different (e.g. a tonal source and a broad band source).

The orientation of the microphone diaphragm to the sound incidence has an influence on the recorded sound especially at higher frequencies. The correction factor $C_{\mathrm{ang},i}$ takes into consideration such influences. As it may be assumed, the correction is the same for the primary and the transfer source if the microphone position is not altered and the correction is apparently not influential in the substitution method.

3.9.2 Corrections to be applied and uncertainty equation

After paragraph 3.9.1 the surface and time averaged sound pressure level for each source becomes:

$$\overline{L_{p,\mathrm{PS}}} = 10\lg\left[\frac{1}{n_{\mathrm{mic}}} 10^{0.1 C_{\mathrm{noise,\,PS}}} \sum_{i=1}^{n_{\mathrm{mic}}} 10^{0.1 L'_{p,\mathrm{PS},i}}\right] \mathrm{dB} \tag{3.6}$$

and

$$\overline{L_{p,\mathrm{TS}}} = 10\lg\left[\frac{1}{n_{\mathrm{mic}}} 10^{0.1\left(C_{\mathrm{noise,\,TS}} + C_{\mathrm{FFT}} + C_{\mathrm{nf}} + C_{\mathrm{scr}}\right)} \sum_{i=1}^{n_{\mathrm{mic}}} 10^{0.1 L'_{p,\mathrm{TS},i}}\right] \mathrm{dB} \tag{3.7}$$

The expression of the uncertainty follows the definition of GUM [JCG08], according to which the measurand Y usually cannot be directly measured, but is determined by various input quantities X_i instead:

$$Y = f\left(X_i\right) \tag{3.8}$$

The input quantities of Eq.(3.8) are considered uncorrelated random variables that follow certain distributions. The expected value of the measurand is then determined by the estimates of the input quantities and the combined uncertainty is expressed as:

$$u(y) = \sqrt{\sum_{i=1}^{n_{\mathrm{est}}} \left[\frac{\partial f}{\partial x_i} u\left(x_i\right)\right]^2} \tag{3.9}$$

where n_{est} is the number of estimates.

The correction factors in Eqs.(3.6) and (3.7) can be considered independent of each other. Using Eq.(3.9) the combined uncertainty of each measurand is:

$$u^2\left(\overline{L_{p,\mathrm{PS}}}\right)=\left(\frac{\partial \overline{L_{p,\mathrm{PS}}}}{\partial L'_{p,\mathrm{PS},i}}\right)^2 u^2\left(L'_{p,\mathrm{PS},i}\right)+\left(\frac{\partial \overline{L_{p,\mathrm{PS}}}}{\partial C_{\mathrm{noise,\,PS}}}\right)^2 u^2\left(C_{\mathrm{noise,\,PS}}\right) \tag{3.10}$$

$$u^2\left(\overline{L_{p,\mathrm{TS}}}\right)=\left(\frac{\partial \overline{L_{p,\mathrm{TS}}}}{\partial L'_{p,\mathrm{TS},i}}\right)^2 u^2\left(L'_{p,\mathrm{TS},i}\right)+\left(\frac{\partial \overline{L_{p,\mathrm{TS}}}}{\partial C_{\mathrm{noise,\,TS}}}\right)^2 u^2\left(C_{\mathrm{noise,\,TS}}\right) + \left(\frac{\partial \overline{L_{p,\mathrm{TS}}}}{\partial C_{\mathrm{FFT}}}\right)^2 u^2\left(C_{\mathrm{FFT}}\right)+\left(\frac{\partial \overline{L_{p,\mathrm{TS}}}}{\partial C_{\mathrm{nf}}}\right)^2 u^2\left(C_{\mathrm{nf}}\right)+\left(\frac{\partial \overline{L_{p,\mathrm{TS}}}}{\partial C_{\mathrm{scr}}}\right)^2 u^2\left(C_{\mathrm{scr}}\right) \tag{3.11}$$

The determination of the sensitivity coefficients yields the following expressions for the combined uncertainty:

$$u^2\left(\overline{L_{p,\mathrm{PS}}}\right)=u^2\left(\overline{L'_{p,\mathrm{PS}}}\right)+u^2\left(C_{\mathrm{noise,\,PS}}\right) \tag{3.12}$$

$$u^2\left(\overline{L_{p,\mathrm{TS}}}\right)=u^2\left(\overline{L'_{p,\mathrm{TS}}}\right)+u^2\left(C_{\mathrm{noise,\,TS}}\right)+u^2\left(C_{\mathrm{FFT}}\right)+u^2\left(C_{\mathrm{nf}}\right)+u^2\left(C_{\mathrm{scr}}\right) \tag{3.13}$$

3.9.3 Uncertainty due to background noise

The correction factor for the background noise C_{noise} is given by ISO6926 [ISO6926]:

$$C_{\mathrm{noise}}=-10\lg\left[1-10^{-0.1\left(\overline{L'_p}-\overline{L_B}\right)}\right]\mathrm{dB} \tag{3.14}$$

where $\overline{L_B}$ is the average background noise level of all microphones. The uncertainty is calculated according to Eq.(3.9) as:

$$u^2\left(C_{\mathrm{noise}}\right)=\left[\frac{10^{-0.1\left(\overline{L'_p}-\overline{L_B}\right)}}{1-10^{-0.1\left(\overline{L'_p}-\overline{L_B}\right)}}\right]^2\left[u^2\left(\overline{L'_p}\right)+u^2\left(\overline{L_B}\right)\right] \tag{3.15}$$

where the uncertainties in the second bracket can be calculated by:

$$u^2\left(\overline{L'_p}\right)=\sigma^2\left(\overline{L'_p}\right) \tag{3.16}$$

The standard deviation includes all measurements performed (various radii) referred to the same radius. Similar equation can be applied for the background noise level uncertainty.

Eq.(3.15) shows that the background noise uncertainty depends on the difference between the average sound pressure levels of the source signal and the noise and the uncertainties of these quantities. The background noise uncertainty must be individually calculated for each source included in the substitution method, due to the differences in the spectral content and frequency response of the sources.

3.9.4 Uncertainty due to FFT windowing

In section 3.3 the correction factor due to FFT window C_{FFT} was described. Measurements of a RSS were performed using both uniform and Hanning window. The related uncertainty was calculated as:

$$u^2_{\mathrm{FFT}} = \frac{\sigma^2\left(L_{p,\,\mathrm{han}\,,\,i} - L_{p\,,\,\mathrm{uni}\,,\,i}\right)}{n_{\mathrm{mic}}} \qquad (3.17)$$

where σ is the standard deviation of the sound pressure level differences and n_{mic} the number of microphones.

3.9.5 Uncertainty due to near field effects

The RSS is a dipole based on the sound generation mechanism and a correction for near field effects C_{nf} must be taken into account for the transfer standard calibration procedure. Near field measurements and a theoretical correction are presented in section 4.4. The uncertainty for the theoretical near field correction can be calculated following the regression error analysis described by Eq.(4.24). The theoretical correction for the near field effects does not apply to the room modes which also affect the measurements. For this reason, a comparison between the aforementioned theoretical correction and a correction based on the sound pressure measurements was performed and it was found that the latter also included the room influences and therefore, should be applied to the calibration procedure. Based on this correction an uncertainty is proposed and described in Section 3.9.7.

3.9.6 Uncertainty due to microphone windscreens

The wind produced by the rotating fan of the RSS is a disturbing factor especially at low frequencies and for short measurement radii. For this reason, windscreens may be applied for the RSS measurements. Apart from the desirable wind suppression at low frequencies there is also a high frequency effect with the absorption provided by the porous windscreens. This induces a correction factor C_{scr} and a related uncertainty. Due to the same exposure of each microphone to the high frequency effect, the correction is globally applied to the time and surface averaged sound pressure level. The correction factor was experimentally determined by measurements of different RSSs at various radii. At each radius, measurements with and without windscreens were performed. The correction factor was calculated by:

$$C_{\text{scr}} = \overline{\overline{L_{p,\text{TS, scr}}} - \overline{L_{p,\text{TS, no scr}}}} \tag{3.18}$$

The related uncertainty is then:

$$u^2\left(C_{\text{scr}}\right) = \sigma^2\left(C_{\text{scr}}\right) \tag{3.19}$$

3.9.7 Application of the substitution method under calibration conditions

After applying the previously mentioned corrections for the calculation of the surface and time averaged sound pressure levels as described in Eqs.(3.6) and (3.7), the sound power level of the transfer source under calibration conditions can be determined by the sound power level of the primary source as:

$$\begin{aligned} L_{W,\text{TS, cal}} &= L_{W,\text{PS}} + \overline{L_{p,\text{TS, cal}}} - \overline{L_{p,\text{PS}}} \\ &+ \left[-10\lg\left(\frac{B_{\text{TS, cal}}}{B_{\text{PS}}}\right)\text{dB} + 5\lg\left(\frac{T_{\text{TS, cal}}}{T_{\text{PS}}}\right)\text{dB}\right] + C_{3,\text{TS, cal}} - C_{3,\text{PS}} \end{aligned} \tag{3.20}$$

The term in parenthesis is the reference quantity correction and C_3 the correction for air absorption as described in ISO 3745 [ISO3745] and ISO 9613-1 [ISO9613]. The correction for the reference meteorological conditions (C_2) has not been included, because it lies out of the scope of this study, where the reference is the free field sound power level of the primary standard.

The uncertainty of the sound power level is given by:

$$
\begin{aligned}
u^2\left(L_{W,\mathrm{TS,cal}}\right) = {} & u^2\left(L_{W,\mathrm{PS}}\right) + \sigma^2\left(\overline{L_{p,\mathrm{TS,cal}}} - \overline{L_{p,\mathrm{PS}}}\right) \\
& + \sum_{i=1}^{n_r}\left[\overline{\left(\overline{L_{p,\mathrm{TS,cal}}} - \overline{L_{p,\mathrm{PS}}}\right)_{r,i}} - \overline{\left(\overline{L_{p,\mathrm{TS,cal}}} - \overline{L_{p,\mathrm{PS}}}\right)}^{r}\right]^2 \\
& + \left[\frac{10\,\mathrm{dB}}{\ln 10}\frac{B_{\mathrm{TS,cal}}}{B_{\mathrm{PS}}}u\left(\frac{B_{\mathrm{TS,cal}}}{B_{\mathrm{PS}}}\right)\right]^2 + \left[\frac{5\,\mathrm{dB}}{\ln 10}\frac{T_{\mathrm{TS,cal}}}{T_{\mathrm{PS}}}u\left(\frac{T_{\mathrm{TS,cal}}}{T_{\mathrm{PS}}}\right)\right]^2 \\
& + u^2\left(\overline{L'_{p,\mathrm{PS}}}\right) + u^2\left(\overline{L'_{p,\mathrm{TS}}}\right) + u^2\left(C_{\mathrm{noise,PS}}\right) + u^2\left(C_{\mathrm{noise,TS}}\right) \\
& + u^2\left(C_{3,\mathrm{TS,cal}} - C_{3,\mathrm{PS}}\right) + u^2\left(C_{\mathrm{FFT}}\right) + u^2\left(C_{\mathrm{scr}}\right)
\end{aligned}
\tag{3.21}
$$

where n_r is the number of the different radii at which measurements were performed.

3.9.8 Uncertainty of the primary source sound power level

The sound power level of the primary source is determined by vibration velocity measurements performed by a laser scanning vibrometer. For further information on the primary source the reader is kindly requested to read [Kir06]. The uncertainty of the sound power of the primary source has been provided by relevant measurements focusing on the realization of unit watt in airborne sound.

3.9.9 Uncertainty due to the surface and time averaged sound pressure level difference

The substitution method includes the sound pressure level difference of the primary and transfer source. The uncertainty calculation must include a statistical effect and any systematic effects, such as near field effects and remaining room influences. The former is expressed as the standard deviation of all sound pressure level differences and the latter by deviations between the mean value for a measurement distance and the mean value for all radii. Since the systematic effects are related to low frequencies, the deviations were set to zero for frequencies above 1 kHz.

3.9.10 Uncertainty due to reference quantity correction

The reference quantity correction has been expressed by including the atmospheric pressure and ambient temperature variations during the primary and transfer source measurements. The ratio of atmospheric pressures and ambient temperatures for the measurements performed at the same location is in the range:

$$\frac{B_{\mathrm{TS,\,cal}}}{B_{\mathrm{PS}}} = 0.99...1.02, \quad \frac{T_{\mathrm{TS,\,cal}}}{T_{\mathrm{PS}}} = 0.99...1.01 \tag{3.22}$$

The uncertainty estimation following Eq.(3.21) for these ratios has been calculated to be 0.016 dB for atmospheric pressure and 0.002 dB for ambient temperature variations. The uncertainty for the measurement of both ratios is estimated to be 0.1%, because the measurements were performed by the same instruments. Apparently, the uncertainty contribution is infinitesimal.

Measurements were performed at three different radii (1.45 m, 1.70 m & 2 m) including three transfer sources and the primary source. Five measurements were made pro radius providing 15 surface and time averaged sound pressure levels for each source.

3.9.11 Uncertainty due to air absorption correction

If the measurement of the transfer source is directly followed by this of the primary source and the changes in atmospheric pressure, ambient temperature and relative humidity are small, the correction can be neglected. As it has been expressed by Eq.(3.20), the substitution method includes the difference of the corrections and apparently, the uncertainty must be calculated based on the difference as well. For this study the variations as expressed by Eq.(3.22) and including measurements on different dates, lead to a maximum difference between the air absorption correction for the primary and the transfer source of 0.27dB. The related uncertainty can be calculated by:

$$u^2\left(C_{3,\,\mathrm{TS,\,cal}} - C_{3,\,\mathrm{PS}}\right) = \sigma^2\left(C_{3,\,\mathrm{TS,\,cal}} - C_{3,\,\mathrm{PS}}\right) \tag{3.23}$$

An alternative way to calculate the uncertainty would be to apply Eq.(3.9) to the air correction equations described in ISO 9613-1 [ISO9613]. It is though believed that the number of the available measurements on different environmental conditions is sufficient to provide the wanted uncertainty.

Another uncertainty contributing factor may be the difference of the acoustic centre location between the sources. The transfer sources have a height and thus, an elevated acoustic centre compared to the centre of the primary source, which is on the floor. This is related to distance between the source and the microphone for the air absorption correction. An investigation was performed by calculating the absorption correction for different acoustic centre locations: on the base, in the middle and on top of the transfer source. The related uncertainties were estimated and found to be close to the value provided by Eq.(3.23). This means that the uncertainty contribution due to the location of the acoustic centre can be neglected.

3.9.12 Combined uncertainty for the substitution method under calibration conditions

By taking into consideration the individual uncertainties mentioned above, the combined uncertainty for the first part of the calibration procedure of transfer standards based on sound pressure measurements can be estimated. It must be noted, that the calibration procedure includes further measurements using the transfer standards in realistic environments, for which another application of the substitution method is required in order to relate the sound power level of the transfer standard under calibration conditions to the in situ sound power level. This is explicitly described in the following chapter.

The combined uncertainty for the first part of the calibration procedure is:

$$\begin{aligned} u^2\left(L_{W,\mathrm{TS,cal}}\right) &= u^2\left(L_{W,\mathrm{PS}}\right)+u^2\left(\overline{L_{p,\mathrm{TS,cal}}}-\overline{L_{p,\mathrm{PS}}}\right)+u^2\left(\overline{L'_{p,\mathrm{PS}}}\right)+u^2\left(\overline{L'_{p,\mathrm{TS}}}\right) \\ &+u^2\left(C_{\mathrm{noise,PS}}\right)+u^2\left(C_{\mathrm{noise,TS}}\right)+u^2\left(C_{\mathrm{scr}}\right)+u^2\left(C_{3,\mathrm{TS,cal}}-C_{3,\mathrm{PS}}\right)+u^2\left(C_{\mathrm{FFT}}\right) \end{aligned} \tag{3.24}$$

The combined uncertainty was calculated for three different RSSs (B&K, EDF, NOR) individually and it was found that there are no significant uncertainty differences between the sources. For this reason, a global combined uncertainty was calculated to cover all measured RSSs. To achieve this, the values of each uncertainty component were compared for the three RSSs and the maximum value was chosen for the global uncertainty calculation, which is presented in Figure 3.11 and Figure 3.12. The total uncertainty and its components for each RSS are shown in Appendix B. In one-third octave bands, at low frequencies the uncertainty is influenced by the deviation of the sound pressure level differences, near field effects, remaining room influences and the influence of the background noise for the primary source. At high frequencies, the deviation of the sound pressure level differences, the deviation of the primary source and transfer source sound pressure level and the windscreen uncertainty become influential. The same applies to FFT analysis with the high frequencies being also influenced by the primary source background noise uncertainty.

The next step of the dissemination of the unit watt is the sound power determination of the transfer standard in situ, which is discussed in the following chapter. The transition from calibration conditions to in situ conditions presupposes the correction of the sound power level. The factors that affect the sound power is atmospheric pressure, ambient temperature and fan rotation speed. The related correction factors were derived after corresponding measurements and the overall correction is presented along with its uncertainty. Additionally, the proposed correction is compared to an existing one.

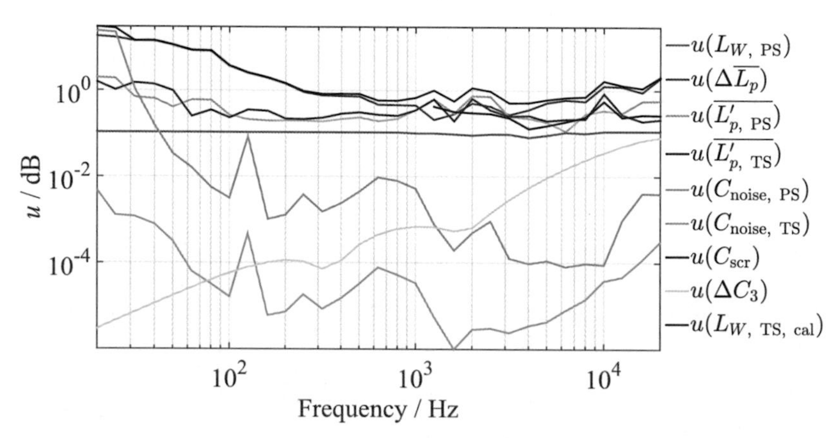

Figure 3.11: Combined and individual uncertainty for the transfer standard sound power determination under calibration conditions for one-third octave bands.

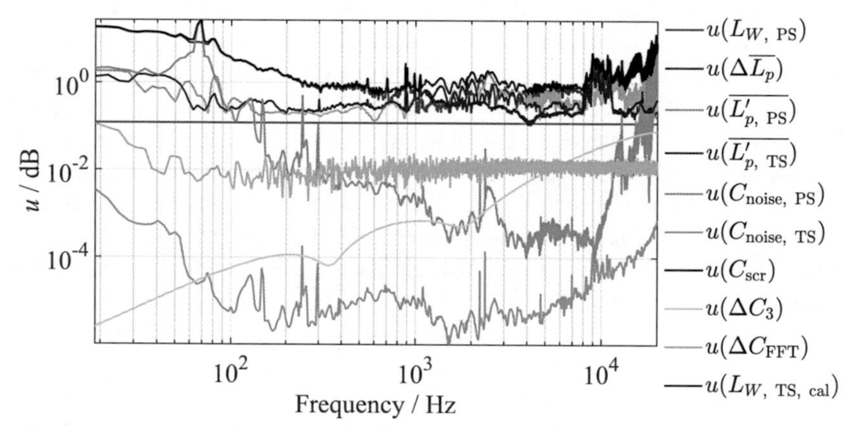

Figure 3.12: Combined and individual uncertainty for the transfer standard sound power determination under calibration conditions for FFT frequency bands.

4 Properties of transfer standards

For the determination of sound power by applying the substitution method aerodynamic reference sound sources may be used, which must fulfil the requirements proposed by the relevant ISO [ISO6926]. The types depicted in Figure 3.1 were investigated as potential candidates for transfer standards in the dissemination process. In this chapter the results of these investigation are presented.

4.1 Existing requirements

The current performance requirements, according to ISO 6926 [ISO6926] include the temporal stability of the sound power output, the total sound power output, the spectral characteristics, the directivity and the recalibration. The temporal stability is quantified by the standard deviation of the sound power level under repeatability conditions and three values are proposed for three different frequency regions as shown in Table 4.1.

Table 4.1: Maximum value of the standard deviation under repeatability conditions of the sound power level output of an aerodynamic reference sound source.

Frequency range (Hz)	**Standard deviation under repeatability conditions σ_r (dB)**
50 – 80	0.8
100 – 160	0.4
200 – 10000	0.2

The spectral characteristics must be broadband (octave or one-third octave) for the frequency range of use. Although narrow band frequency analysis is not taken into consideration in the qualification procedure, such analysis was also performed since sources with tonal frequency characteristics are aimed to be included in the dissemination process.

The directivity is quantified by the directivity index (DI), which is calculated by:

$$D_{Ii} = L_{pi} - \overline{L_p} \tag{4.1}$$

where L_{pi} is the sound pressure level recorded by the i-th microphone used for the determination of the sound power and $\overline{L_p}$ is the surface and time averaged sound pressure level, both in dB. The maximum allowable directivity index value is +6 dB [ISO6926].

4.2 Temporal stability

For the results of this chapter six RSSs were used, an AEG, three Brüel & Kjær, an EDF and a Norsonic. Each RSS under measurement was placed at the centre of PTB's hemianechoic room and the emitted sound was recorded using the scanning apparatus as described in chapter 3.1. The covered measurement surface had three different radii (1.45, 1.70 & 2m). For each source, four measurements were performed per radius in a six-month period. Apart from the scanning apparatus, a stationary quarter circle metallic arc of 2m radius was also used. Along its body ten 1/4" condenser free field microphones (G.R.A.S. Type 40 BF) were positioned as seen in Figure 4.1. The elevation angles from the hemianechoic room floor were: 3°, 9°, 16°, 23°, 30°, 37°, 46°, 56°, 66° and 90°. The position was chosen that the measurement surface covered by each microphone is the same. The stationary arc measurements were performed during a two-month period and were grouped as eleven sets pro source (each set measured on different date). Prior to measurements the microphones were calibrated using a pistonphone (Brüel & Kjær Type 4228). Each measurement set consisted of ten consecutive measurements of the same source. For each RSS, a total of 110 measurements became available except for one, where due to prior to warm up instabilities, the first measurement of each set was discarded leading to a total of 99 measurements. For the investigation of the temporal stability in narrow band frequency analysis, apart from the standard octave or one-third octave band analysis [ISO6926] an FFT analysis (3.125 Hz resolution, 6401 lines, Hanning window) was also applied. The stationary arc measurements were performed real time whereas the scanning apparatus measurements included post analysis of the recorded wav files.

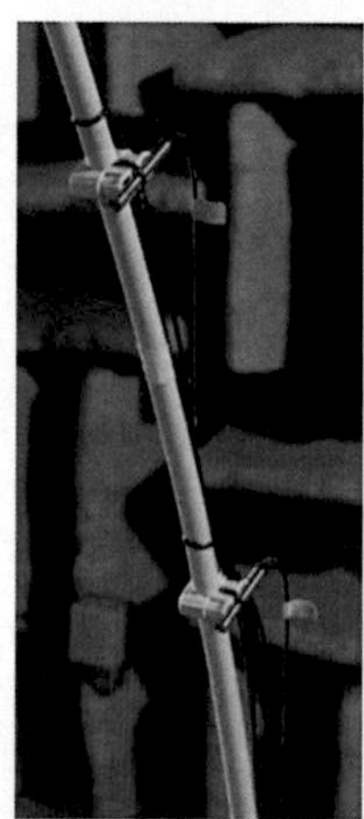

Figure 4.1: Quarter-circular stationary arc for the investigation of reference sound source temporal stability (left) and two zoomed microphones (right).

An important setting for the temporal stability measurements was the measurement duration, especially for the narrow band resolution. The standard deviation σ of a random signal is given by:

$$\sigma = \frac{4.34}{\sqrt{B_{\text{band}} T_{\text{dur}}}} \text{dB} \tag{4.2}$$

where B_{band} is the signal bandwidth in Hz and T_{dur} the signal duration in s [Bie03].

The measurement duration for the stationary arc measurements was set to 610 s, which according to Eq.(4.2) corresponds to a standard deviation of 0.1 dB. The scanning measurements had 1200 s duration ($\sigma = 0.07$ dB). The randomness of the RSS signal was also checked by varying the measurement time while keeping the bandwidth constant. Shorter measurement durations led to higher standard deviation.

The sound power level was determined by:

$$L_W = 10 \lg \left(\sum_i \frac{S_i}{S_0} \times 10^{L_{pi}/10} \right) \text{dB} - 10 \lg \left(\frac{\rho c}{400} \frac{\text{m}^3}{\text{kg} \cdot \text{s}} \right) \text{dB} \tag{4.3}$$

where L_{pi} is the sound pressure level of the i-th microphone and S_i the surface covered by the i-th microphone in m^2 [Ver06]. The sound pressure level has been not corrected for background noise because the stationary signal of the RSSs cannot be fully corrected for the non-stationary background noise. On the contrary, the correction for the air absorption was applied according to ISO 6926 [ISO6926].

Figure 4.2 shows the sound power levels of the Norsonic RSS for both broad and narrow band analysis after the stationary arc measurements. As it can be seen, the emitted sound contains tonal components, which are revealed by narrow band analysis.

The sound power levels of each source were divided to four groups, a group per measurement radius for the scanning apparatus measurements and a group containing all levels for the stationary arc measurements. This was performed because the reflections from the arc would increase the overall standard deviation in case the sound power levels were taken into consideration as a single set of measurement data, since for each measurement radius the reflection effects are apparent at different frequencies. The repeatability standard deviation was calculated for each RSS and was compared to the proposed values according to Table 4.1. Due to the lack of standard deviation limit values for narrow band analysis, the broad band values were applied to the narrow band measurement results. The temporal stability results are presented in Figure 4.3.

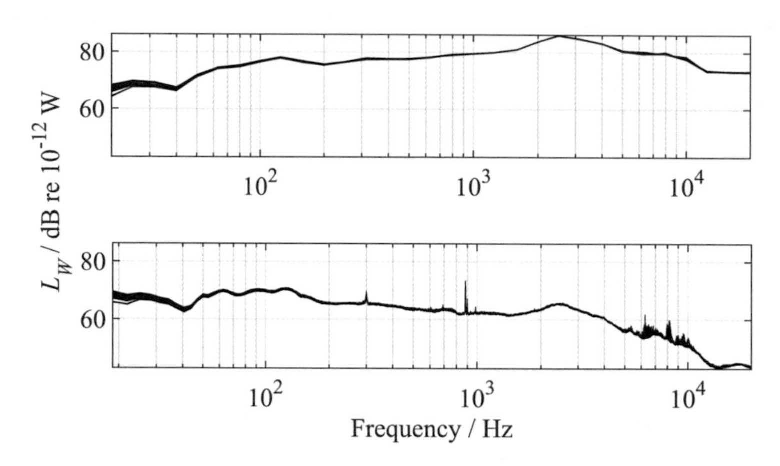

Figure 4.2: Sound power level of the NOR reference sound source for one-third octave bands (top) and FFT bands (bottom).

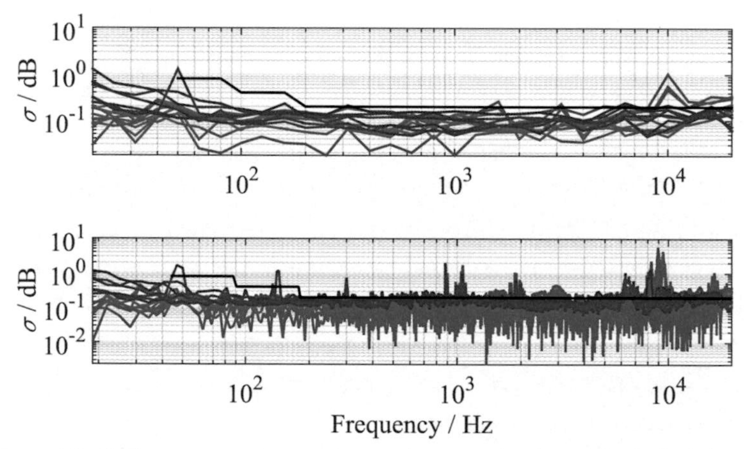

Figure 4.3: Reference sound source sound power level standard deviation under repeatability conditions for one-third octave bands (top) and FFT bands (bottom). Blue: stationary arc measurements. Orange red: scanning apparatus measurements. Black: ISO 6926 limits.

The standard requirements [ISO6926] are fulfilled by almost all RSSs under consideration in case of broadband frequency analysis. A RSS exceeds the ISO limits at low frequencies, which is related to the not constant fan blade passing frequency.

Similarly, only one RSS is related to standard deviation values above limits at high frequencies. This may be attributed to the tonal components at this frequency range, which have a varying level as seen also in FFT analysis. It is also of interest that the deviation is larger in the case of the scanning apparatus measurements than the stationary arc measurements. Concerning the FFT analysis, it can be seen that for frequencies above 200 Hz more sources do not fulfil the repeatability requirements and this becomes more intense above 5 kHz. The 3.125 Hz FFT frequency resolution data was used for the calculation of the standard deviation for different frequency resolutions (6.25, 12.5 and 25 Hz). Figure 4.4 shows the percentage of the signal samples, which fulfil the ISO requirements [ISO6926]. As it can be seen, the percentage increases as the frequency analysis becomes broader. Based on the above, if the qualification of the RSSs is to include narrow band frequency analysis, corresponding standard deviation values must be set.

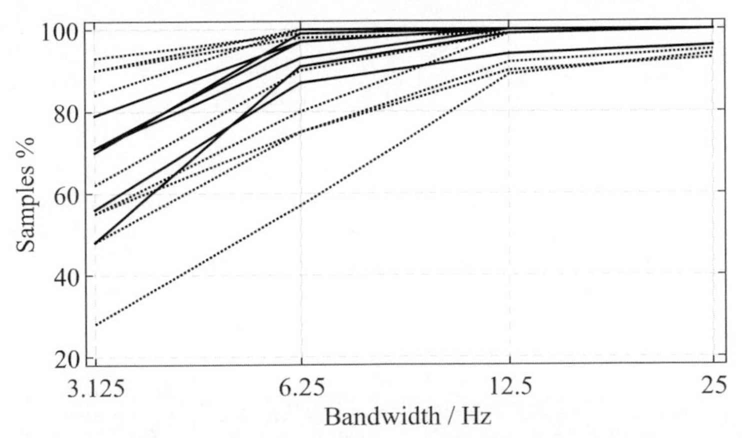

Figure 4.4: Percentage of the samples, which fulfil the ISO 6926 broad band repeatability requirements for various bandwidths. Continuous: stationary arc measurements. Dotted: scanning apparatus measurements.

4.3 Corrections for changes in environmental and operational conditions

The sound power of an aerodynamic RSS (a rotating fan) depends on three main parameters: atmospheric pressure, ambient temperature and fan rotation speed ([Hü772], [Hü981], [Wit02]). This means a correction needs to be applied in case the determination of the sound power of such a source takes place under different environmental and operational conditions. During dissemination, the sound power level of a RSS is to be determined firstly under calibration conditions (in a hemianechoic room) and secondly, in situ (in a realistic environment). In general, an overall correction factor must be included and this may be expressed by:

$$L_{W,\,\text{in situ}} = L_{W,\,\text{cal}} + C \tag{4.4}$$

where C is the overall applied correction.

The following paragraphs describe the results concerning the investigation on the influence of the above parameters on sound generation of a RSS.

4.3.1 Influence of atmospheric pressure

Following the analysis of Wittstock ([Wit02], [Hü981]) the in situ radiated sound power of a fan is related to the sound power of the same fan under calibration conditions according to:

$$P_{\text{in situ}} = P_{\text{cal}} \left(\frac{B_{\text{in situ}}}{B_{\text{cal}}}\right)^{n_B} \left(\frac{T_{\text{in situ}}}{T_{\text{cal}}}\right)^{n_T} \left(\frac{\omega_{\text{in situ}}}{\omega_{\text{cal}}}\right)^{n_\omega} \tag{4.5}$$

where P is the sound power in W, B is the atmospheric pressure in kPa, T is the ambient temperature in K and ω is the fan rotation speed in Hz. The factors n_B , n_T and n_ω are determined by the emission characteristics of the source. By combing Eq.(4.4) and (4.5), the relation between the sound power under calibration conditions and in situ for a RSS can be described by:

$$L_{W,\,\text{in situ}} = \\ L_{W,\,\text{cal}} + 10 n_B \lg\left(\frac{B_{\text{in situ}}}{B_{\text{cal}}}\right) \text{dB} + 10 n_T \lg\left(\frac{T_{\text{in situ}}}{T_{\text{cal}}}\right) \text{dB} + 10 n_\omega \lg\left(\frac{\omega_{\text{in situ}}}{\omega_{\text{cal}}}\right) \text{dB} \tag{4.6}$$

For the investigation of atmospheric pressure influence, measurements were performed in PTB's hemianechoic room at different atmospheric pressure values (97.52 kPa to 102.51 kPa), while the variations in the other two conditions were negligible (ambient temperature ±0.4% and fan rotation speed did not exceed ±1.2%). The stationary arc shown in Figure 4.1 was used for the measurements. The sound power levels were calculated using Eq.(4.3) by also applying air the absorption correction [ISO6926]. The mean value of the ambient temperature and fan rotation speed values were used as the calibration values in Eq.(4.6). The proposed correction was applied using theoretical n factors values ($n_{B,\,\text{theo}} = 1$, $n_{T,\,\text{theo}} = -2.5$ and $n_{\omega,\,\text{theo}} = 5.5$) ([Wit02], [Hü981]). The sound power levels corrected for the influence of the ambient temperature and fan rotation speed were then used for the calculation of n_B. The sound power levels of the RSSs are shown in Figure C.1-Figure C.6 in Appendix C. As it can be seen in Eq.(4.6) the factor may be calculated by applying a regression analysis to the sound power level difference as a function of the atmospheric pressure ratio. A least squares fit was applied for the calculation of all n factors.

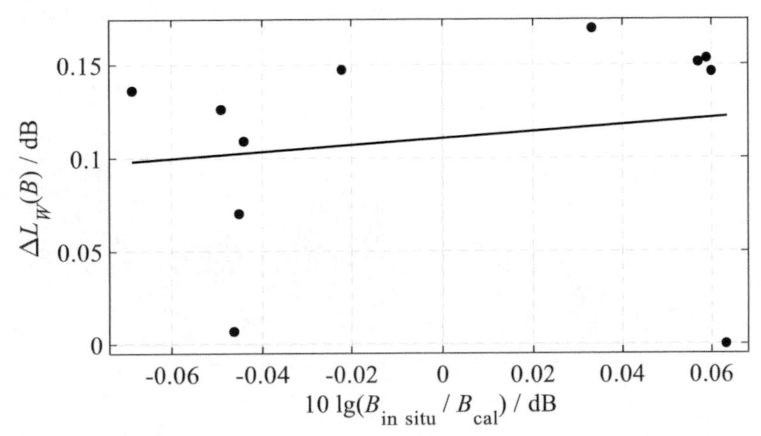

Figure 4.5: Regression analysis for the calculation of the sound power correction factor for influences due to atmospheric pressure changes for a given one-third octave band.

Figure 4.5 shows an example of the regression analysis applied for the calculation of n_B with

$$\Delta L_W(B) = L_W\left(B_{\text{in situ},i}\right) - L_W\left(B_{\text{in situ},\max}\right) \tag{4.7}$$

where $L_W\left(B_{\text{in situ},i}\right)$ is the sound power level measured at the i-th atmospheric pressure and $L_W\left(B_{\text{in situ},\max}\right)$ the sound power level measured at the maximum atmospheric pressure.

For the calculation, all RSSs presented in chapter 4.2 were used.

Table C.1 summarizes the atmospheric pressure, ambient temperature and fan rotation speed values used for the n_B factors calculation. As it may be assumed, the regression analysis of each source produced n_B values for each RSS. The overall n_B value was calculated as the mean value of all RSS values. The final value is presented in Figure 4.6 for both one-third octave band and FFT analysis.

The correction factor values are more close to zero instead to unity as it would be expected ([Hü981], [Hü011]). This is assumingly attributed to the limited variation of the atmospheric pressure, since they were all performed at the same altitude. It would be advisable for such measurements to be repeated at locations with different altitude. Another argument would be that the small changes in the sound pressure level cannot be observable in view of the measurement repeatability. For the rest calculations including the correction for the changes in atmospheric pressure, the theoretical value was used ($n_{B,\,\mathrm{theo}} = 1$).

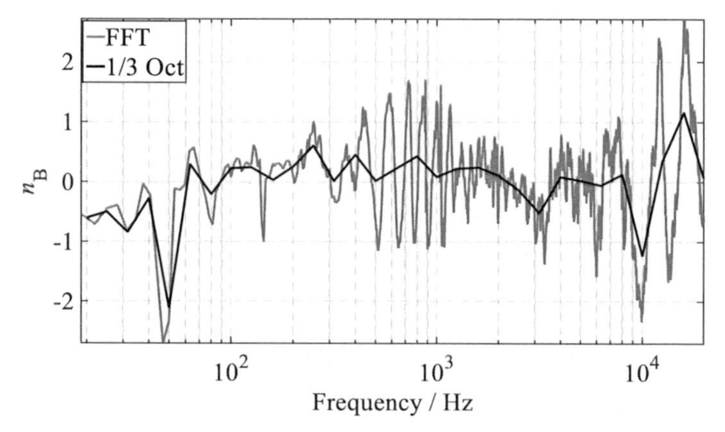

Figure 4.6: Sound power level correction factor for changes in atmospheric pressure.

4.3.2 Influence of ambient temperature

According to ISO 6926 [ISO6926] the sound power under different meteorological conditions must be referred to reference meteorological conditions to compensate for the related radiation impedance variances. On the same theoretical basis and in order to calculate the factor n_T of Eq.(4.6), measurements (indoor and outdoor) were performed at different ambient temperatures.

The outdoor measurements took place in a parking area of PTB with the source located at its centre to avoid reflections from the surrounding buildings. A mobile (for transportation) but stationary (for measurements) semicircular aluminium arc was constructed, along which ten 1/2" condenser free field microphones (G.R.A.S. Type 40 AF) could be positioned. The positioning of the microphones was the same as for the stationary arc of Figure 4.1, related to both sides of the source. The arc for the outdoor measurements is presented in Figure 4.7. The indoor measurements were performed in PTB's hemianechoic room using the same arc.

Three RSSs were used for the measurements, a Brüel & Kjær, the EDF and the Norsonic. The recorded signals consisted of waterfall slices of 1 s duration. The overall signal duration was 1200 s. The originally recorded sound pressure data exhibited large amplitude variations due to high background noise. After the recordings, a statistical analysis of the sound pressure levels was performed, under the assumption that valid levels should not differ more than 25% from the mean sound pressure level (-2.50 dB and 1.94 dB). The statistical analysis led to signal duration between 810 s and 1200 s for one-third octave band analysis and between 321 s and 530 s for FFT analysis (3.125 Hz frequency resolution). The next step in the measurement data evaluation was the background noise correction according to ISO 3745 [ISO3745]. The high levels of the background noise at low frequencies limited the usable frequency range to 315 Hz – 20 kHz. Lastly, an air absorption correction was also applied to the data after noise correction also according to ISO 3745 [ISO3745]. The sound power level was then calculated according to Eq.(4.3).

For the Brüel & Kjær and the Norsonic source, seven measurements were performed (six outdoor and one indoor), while for the EDF source a mechanical failure led to one less outdoor measurement (six in total). Table C.2 contains the operational and environmental parameters during the measurements for the n_T factor calculation.

Figure 4.7: Semicircular stationary arc for outdoor measurements under various ambient temperatures.

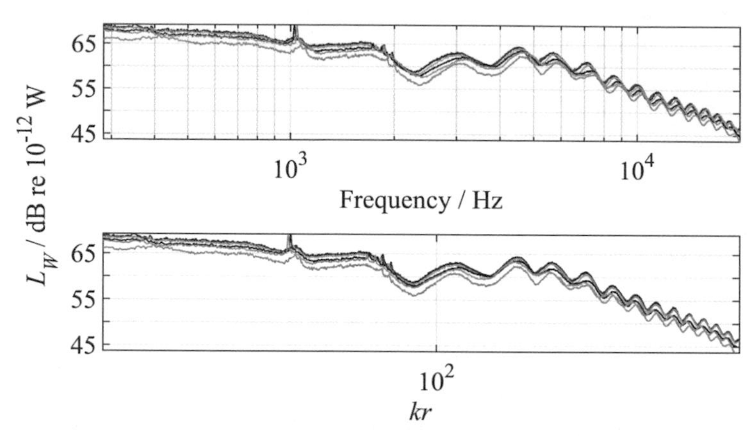

Figure 4.8: Sound power level of the EDF reference sound source at different ambient temperatures against frequency (top) and *kr* (bottom).

As it can be seen in Table C.2, the variation of the atmospheric pressure around mean value was no more than 1% and for the fan rotation speed 1.9%.

The variations in the ambient temperature influenced the original data and especially the FFT data by imposing a high frequency shift. In order to overcome this shift, the data was also examined in relation to *kr*. Figure 4.8 shows the EDF sound power levels for different ambient temperatures plotted against frequency and *kr*. All measured sound power levels are shown in Figure C.7-Figure C.9 in Appendix C. At 2250 Hz the shift in frequency analysis can be seen. This is due to the different ground reflection path between each microphone and the arc for different temperatures. The effect of the temperature is compensated by the *kr* representation. The spectrum ripples are attributed to the reflections from the arc.

The sound power level analysis for the determination of the n_T factor followed the same procedure as described in paragraph 4.2, including both *kr* values and frequency. Initially, *kr* values were calculated for each source, leading to as many *kr* vectors as measurements pro temperature. Then, a global *kr* vector was calculated using linear interpolation of neighbouring samples. This global vector was assumed to represent the frequency range of interest. According to Figure 4.8, the frequency shift due to changes in ambient temperature takes place above 1 kHz. For this reason, the regression was performed for frequency analysis up to 1 kHz and for *kr* analysis for higher frequencies. The same procedure was followed for the uncertainty calculation as well. The missing low frequency part of the spectrum was calculated as the mean value of the rest spectrum for one-third octave bands. The corresponding calculation was performed for each frequency analysis separately, leading to different low frequency values, which comes in contradiction to theoretical expectations. In more, the missing low frequency values were calculated by extrapolation as well. The mean value was finally chosen because it takes into consideration all the available spectrum. The values of n_T are shown in Figure 4.9.

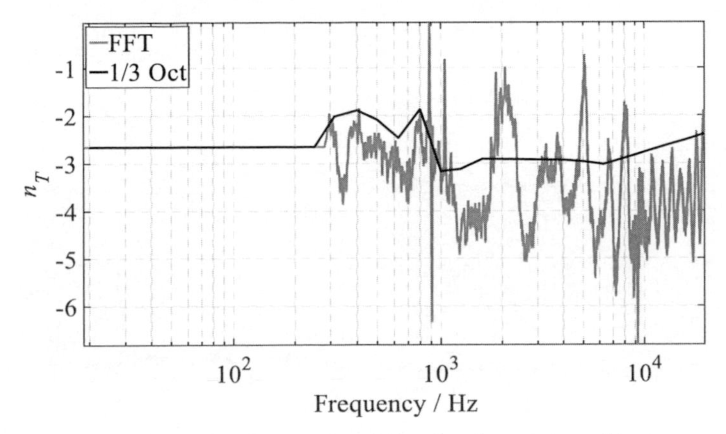

Figure 4.9: Sound power level correction factor for changes in ambient temperature.

According to theory [Wit02] the value of n_T is -2.5 in case of dipole source. As it can be seen in Figure 4.9 the factor value depends on frequency. Ground reflections and reflections from the arc, which could not be fully compensated by the *kr* representation may be observed, especially in FFT analysis. Apparently, the dipole behaviour of the RSSs in all frequency range of interest cannot be supported by Figure 4.9.

4.3.3 Source directivity according to ISO 6926

The three types of RSS used for the influence of ambient temperature were used to investigate the directivity in terms of the DI as described by Eq.(4.1). The data was the same as those used for the scan data of Figure 3.9. Due to the large number of DI values over the measurement surface, only the maximum value of all measurements per source was used and the results are presented in Figure 4.10 for all RSSs under investigation in both broad and narrow band frequency analysis.

The RSSs fulfill the requirements of ISO 6926 [ISO6926] for the one-third octave band analysis. The FFT analysis reveals higher DI values than the corresponding broadband values. The Norsonic RSS has tonal spectral content, which strongly affects the DI above 8 kHz. All three RSSs exhibit close or slightly higher than 6 dB values at 1.6 kHz, especially for the measurements at 2 m measurement radius. This is due to ground reflections for the microphone positions closest to the floor. The effect is decreased for higher positions, but since the maximum DI value is taken into account in Figure 4.10, the reflections effect is more apparent.

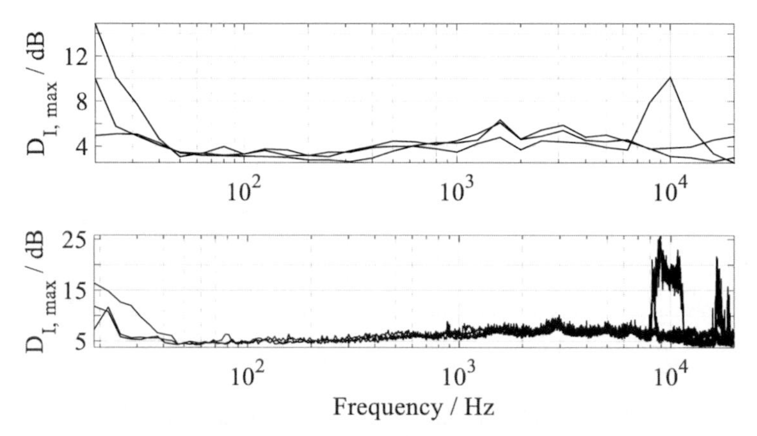

Figure 4.10: Maximum directivity index of the investigated reference sound sources for different measurement radii. Top: one-third octave bands. Bottom: FFT bands.

4.3.4 Source directivity according to spherical harmonics

The radiation pattern of the RSSs may also be analyzed in order to determine the behaviour of the source (e.g. monopole, dipole etc.). This may be achieved by the sound field decomposition of a source using the spherical harmonics transform (SHT). Such analysis is described in the following paragraphs. For calculations the ITA-Toolbox [Ber17] was used (vers.2017, downloaded on 29.08.2017).

4.3.4.1 Spherical harmonics transform

Fourier transform is used to calculate the frequency content of a time domain signal, which is expressed as a sum of sine waves of multiple frequencies. Accordingly, the analysis of a radiation pattern is performed by the SHT ([Hag11], [Sha15]). The basis functions of the SHT are called spherical harmonic functions and they enable the decomposition of the sound field as a sum of spherical harmonics, which have the same meaning as the Fourier transform harmonics.

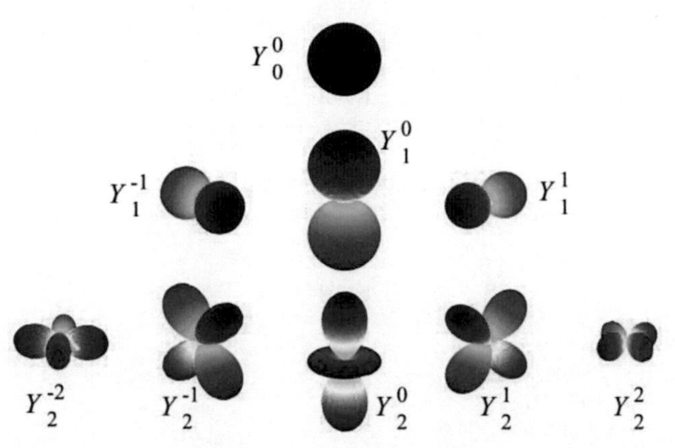

Figure 4.11: First two orders of the real part of the spherical harmonic basis functions.

Consider a the following sound field: a spherical surface that surrounds a source, on which the sound pressure values can be expressed using a spherical coordinate system as shown in Figure 2.2 and described by Eq.(2.1) in the form of $p(k,r,\theta_{\text{el}},\varphi_{\text{az}})$. In the case where the sound field is square-integrable on the surface of the sphere, the sound pressure values on the sphere can be expressed as a spherical harmonics series according to:

$$p(k,r,\theta_{\text{el}},\varphi_{\text{az}})=\sum_{n=0}^{N}\sum_{m=-n}^{n} p_{nm}(k,r)\,Y_n^m(\theta_{\text{el}},\varphi_{\text{az}}) \tag{4.8}$$

The basis functions Y_n^m, most commonly known as the spherical harmonics of order n and degree m, are calculated by:

$$Y_n^m(\theta_{\text{el}},\varphi_{\text{az}}) \triangleq \sqrt{\frac{2n+1}{4\pi}\frac{(n-m)!}{(n+m)!}}\,P_n^m(\cos\theta_{\text{el}})\,e^{jm\varphi_{\text{az}}} \tag{4.9}$$

where P_n^m is the associated Legendre function of order n and degree m.

Figure 4.11 shows the first two orders of the real part of the spherical harmonic functions. The radius represents the magnitude of the functions and the colour the phase.

The spherical harmonics transform of the spatial pressure function can be used to derive the pressure coefficients p_{nm} of Eq.(4.8) according to:

$$p_{nm}(k,r) = \int_{\theta_{\mathrm{el}}=0}^{\pi} \int_{\varphi_{\mathrm{az}}=0}^{2\pi} p(k,r,\theta_{\mathrm{el}},\varphi_{\mathrm{az}})\, Y_n^{m*}(\theta_{\mathrm{el}},\varphi_{\mathrm{az}}) \sin\theta_{\mathrm{el}}\, d\theta_{\mathrm{el}}\, d\varphi_{\mathrm{az}} \tag{4.10}$$

where Y_n^{m*} is the complex conjugate of the spherical harmonics functions.

If the sound pressure coefficients are known through Eq.(4.10), it is possible to calculate the sound pressure at any point outside the sphere, which contains all sources [Pol15].

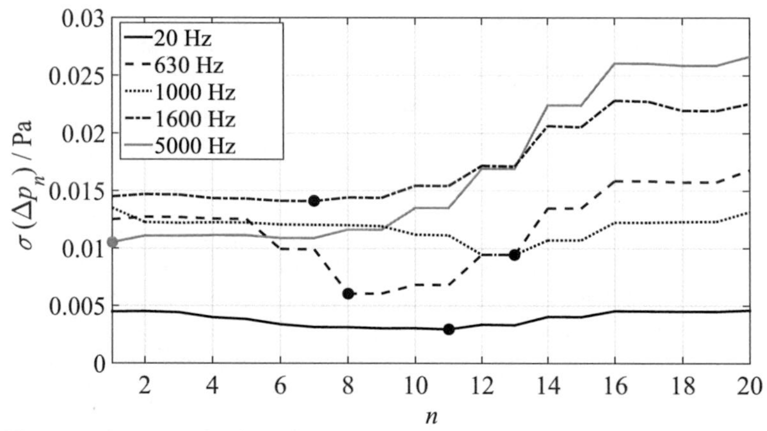

Figure 4.12: Determination of the spherical harmonics order for five one-third octave bands. The selected order is indicated by the dots.

4.3.4.2 Decomposition implementation

The sound pressure measurements were performed in PTB's hemianechoic room using the scanning apparatus described in chapter 3 for three different radii (1.45 m, 1.70 m & 2 m) and the RSSs mentioned at paragraph 4.3.2. Each scan had 1200 s duration and three measurements for each radius were performed, yielding 12 spatial data sets for each source for one-third octave bands. Post analysis of the time recorded signals divided each microphone trajectory to 171 segments. For each segment, the time averaged sound pressure was calculated, providing a hemispherical grid of 4104 points.

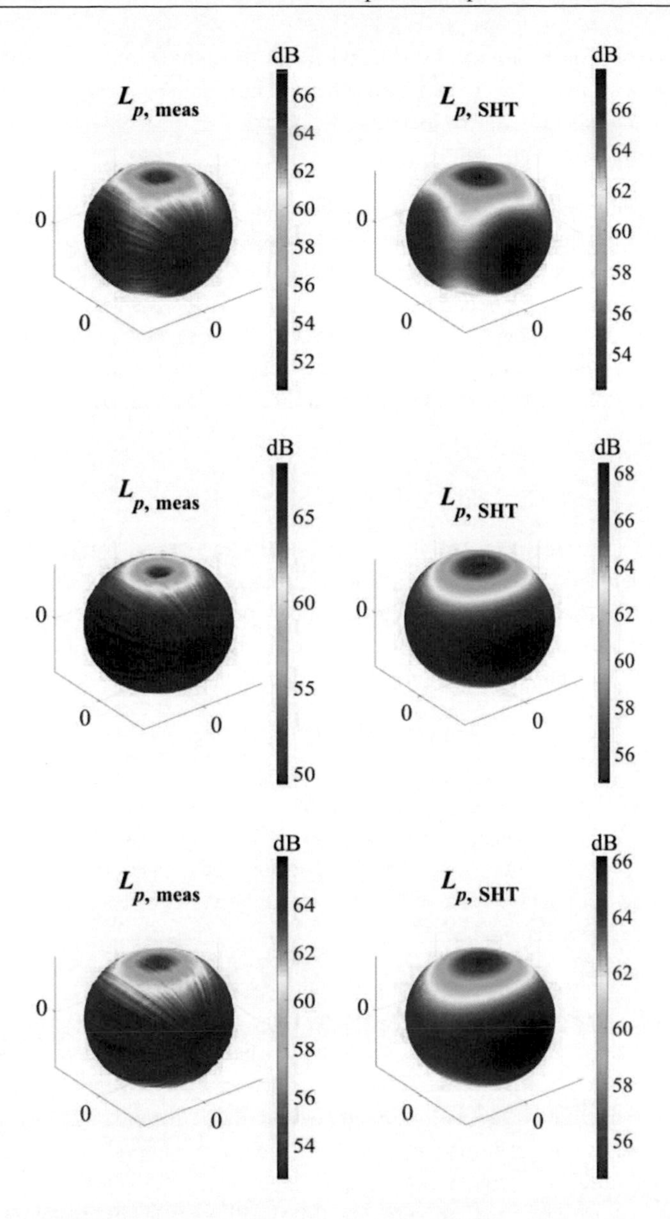

Figure 4.13: Sound pressure level over a sphere for originally measured data (left) and after applying the spherical harmonics transform (right). B&K (top), EDF (middle) and NOR (bottom) source.

Since the previously described spherical harmonics analysis refers to sound pressure measurement over a sphere, the hemispherical measurement grid was used for the creation of a spherical one, by assuming symmetry over the highly reflecting floor of the hemianechoic room.

The originally measured sound pressure data was used for the determination of the spherical harmonics order. Each measured sound pressure set (p_{meas}) was used for the calculation of the sound pressure coefficients p_{nm} using Eq.(4.10) for each order up to a maximum ($n_{\max} = 20$). The sound pressure coefficients were then used for the calculation of new sound pressure values (p_{SHT}) according to Eq.(4.8). The amplitude difference between the original sound pressure and the sound pressure calculated from the spherical harmonics transform was then determined for each order:

$$\Delta p_n = p_{\text{SHT},n} - p_{\text{meas}} \tag{4.11}$$

As it may be assumed, Eq.(4.11) provided as many amplitude differences as measurement grid points. For each measurement the standard deviation of all differences was calculated per frequency and order. The order that corresponded to the minimum standard deviation, was selected for further calculations. Eq.(4.12) summarizes the spherical harmonics order calculation:

$$n(f) = \min\left\{\sigma\left[\Delta p_n(f)\right]\right\} \tag{4.12}$$

Figure 4.12 shows an example of the determination of the spherical harmonics order for five 1/3 octave bands according to Eq.(4.12). Figure 4.13 shows the sound pressure level distribution over a sphere for both the measured and the SHT data for the one-third octave band of 100 Hz. It can be verified that the SHT provided data close to the measured.

Figure 4.14 shows the surface averaged sound pressure level difference between the measured sound pressure and the sound pressure after the SHT for each RSS under consideration. The subtraction was initially applied to all measurement points and the surface averaged value was calculated afterwards. All twelve level differences were close to the average for all RSSs. It must be stated that the same measurements and analysis were performed for the PTB's primary source (a vibrating piston embedded on PTB's hemianechoic room, i.e. a monopole) and the corresponding level difference was close to zero for almost all frequencies, meaning that the determined SHT order was physically correct. By examining Figure 4.14 it can be stated that the average level differences exhibit similarities (maximum value between 1 kHz and 2 kHz) but also differ in the sense of the levels aside the maximum.

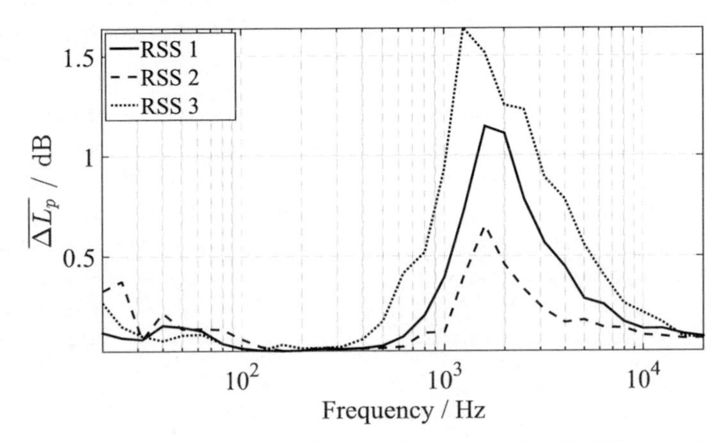

Figure 4.14: Mean surface averaged sound pressure level difference between originally measured data and data after the spherical harmonics transform.

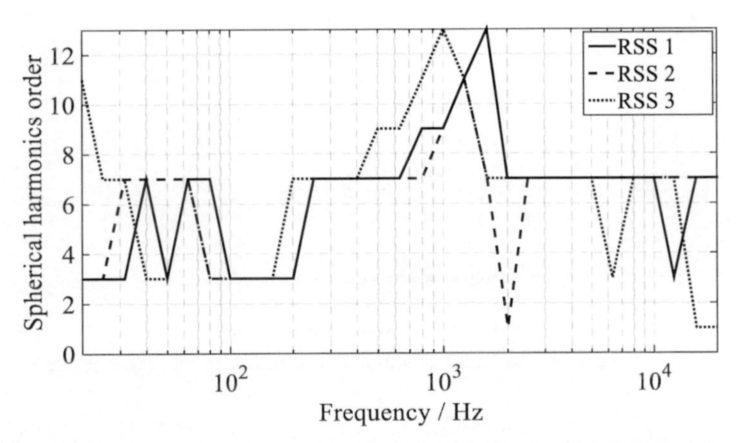

Figure 4.15: Spherical harmonics order for each RSS against frequency.

4.3.4.3 Transfer standard radiation characteristics

The spherical harmonics order was determined as the mean value of the orders corresponding to each measurement set rounded to the nearest integer. Figure 4.15 shows the spherical harmonics order for each RSS. As it can be seen, the spherical harmonics order depends on frequency for all sources, with the highest values between 400 Hz and 2 kHz.

The SHT also enables the calculation of the signal energy distribution to each harmonic. The energy distribution was performed using the ITA-Toolbox [Ber17] as well. An example of the energy distribution for a single frequency (500 Hz) is depicted in Figure 4.16. As it may be seen, the energy is mainly distributed to zeroth, second, fourth and sixth harmonic. In all distributions, the majority of the signal energy is located at the zeroth order.

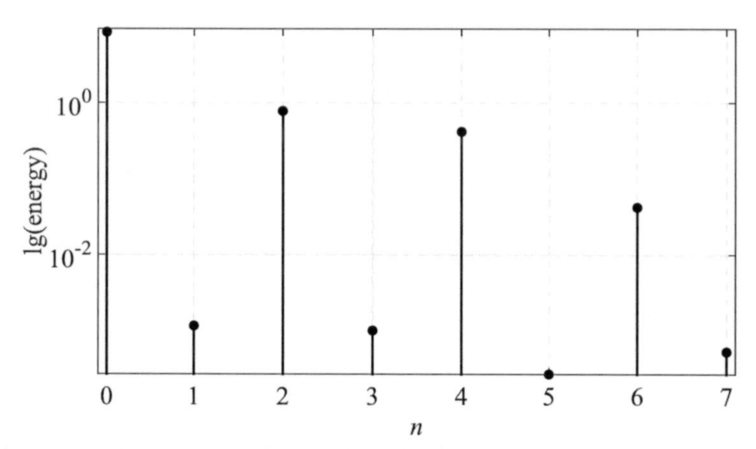

Figure 4.16: Spherical harmonics energy distribution.

4.3.4.4 Comparison to measurement results

The energy distribution from the SHT was used as a tool to relate the radiation characteristics of the RSSs to those calculated in section 4.3.2. To achieve this, a single value would be recommendable to characterize each RSS based on its energy distribution. The expected value of the order was calculated relating the energy of each harmonic to the total according to:

$$n_{\text{expected}} = \sum_{n=0}^{n_{\max}} \frac{nE_n}{E_{\text{tot}}} \tag{4.13}$$

where n is the spherical harmonics order, E_n the energy of each spherical harmonic and E_{tot} the total signal energy.

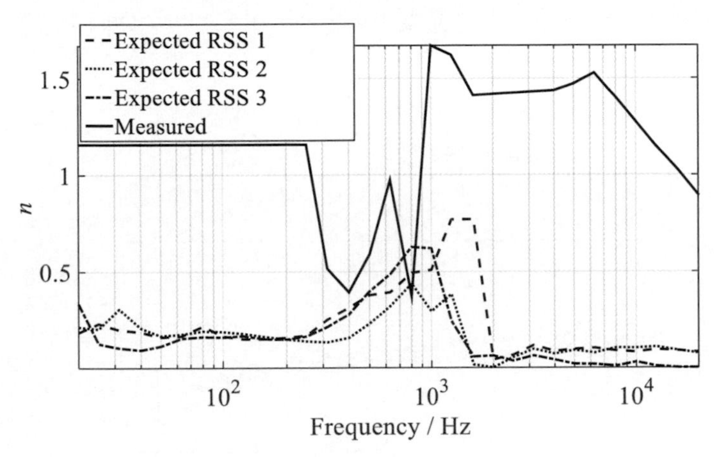

Figure 4.17: Expected spherical harmonics order for each reference sound source (dotted lines) and order based on n_T (continuous line) against frequency.

The factor n_T can be theoretically calculated based on the source radiation order n_{src} (-1 for monopole, -3 for dipole etc.) using:

$$n_T = \frac{n_{\text{src}} - 2}{2} \tag{4.14}$$

n_{src} is related to the spherical harmonics order through:

$$n_{\text{src}} = -2n - 1 \tag{4.15}$$

By combing Eqs. (4.14) and (4.15) the spherical harmonics order can be related to n_T by:

$$n = -n_T - 1.5 \tag{4.16}$$

Figure 4.17 shows the n_{expected} for each RSS along with n having been calculated using Eq.(4.16). The RSS sound production mechanism is related to dipole behaviour. Such behaviour cannot be attributed to the RSSs based on the results of Figure 4.17.

In more, the radiation pattern of a RSS is a combination of more than one source orders with various harmonics energy distribution. Another observation is that the measurements dedicated to the changes in sound power due to ambient temperature changes do not reveal the same source order as the SHT. Apparently, although both analyses (measurement data and SHT) may be used to describe the radiation behaviour of a source, the calculation of n_T can be related to measurements when no directivity information is needed. On the contrary, the directivity of a source may be described by SHT.

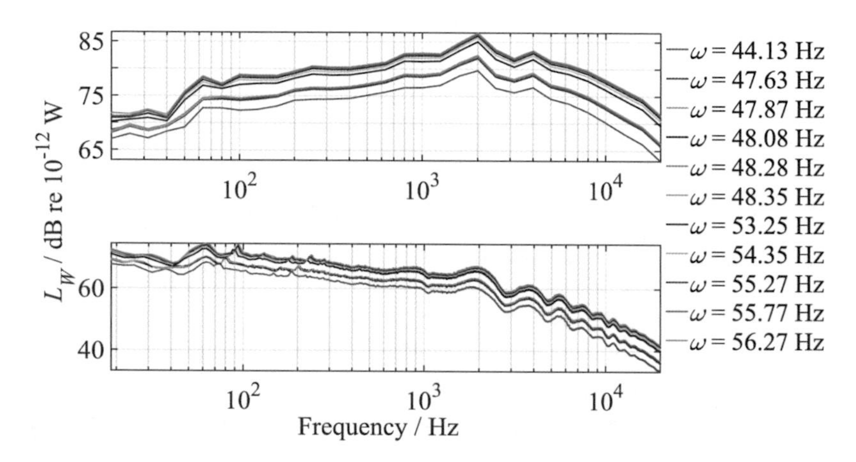

Figure 4.18: Sound power level of the B&K reference sound source for one-third octave bands (top) and FFT bands (bottom) for varying fan rotation speed.

4.3.5 Influence of rotation speed

The variations in the emitted sound power imposed by variations in the fan rotation speed were also studied by varying the RSS input voltage. This was performed using a frequency converter (Philips, Type 2422 530 05405), which also enabled changes in the alternating current frequency (50 Hz and 60 Hz). Three RSSs were initially measured in PTB's hemianechoic room using the scanning apparatus, with only one providing usable sound power variations (Brüel & Kjær). The variations in the input voltage were ± 10 % around the nominal value (230 V for 50 Hz and 115 V for 60 Hz). For the 50 Hz, the input voltage was also decreased by 35 % from the nominal value. Table C.3 shows the environmental and operational values for the calculation of n_ω including the input voltage values. As it can be seen, the variations in the amospheric pressure and ambient temperature were small during measurements.

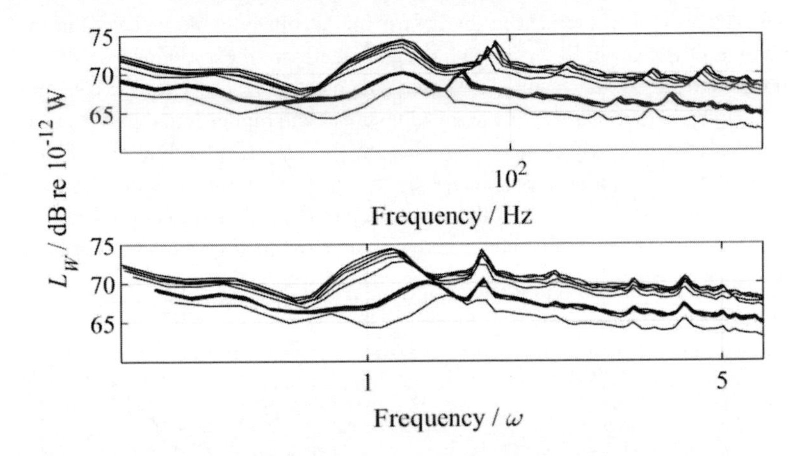

Figure 4.19: Low frequency part of the FFT sound power level of the B&K reference sound source against frequency (top) and frequency over fan rotation speed (bottom).

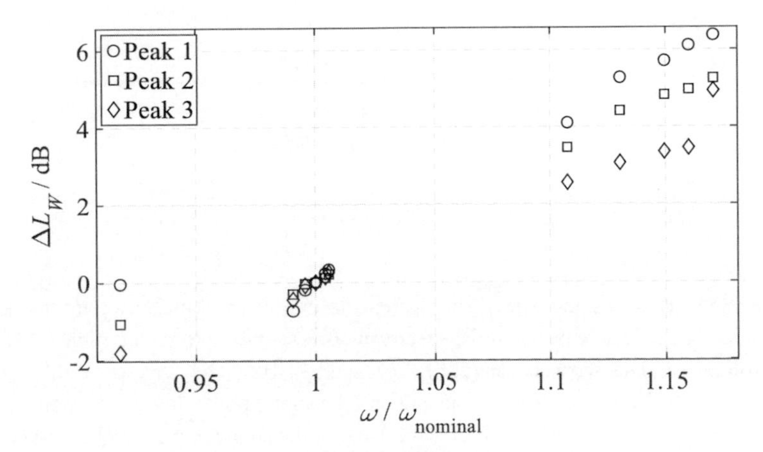

Figure 4.20: Sound power level difference between FFT spectrum peaks at nominal input voltage and rest input voltages.

Figure 4.18 shows the measured sound power levels after having applied air absorption correction [ISO3745]. Compared to the environmental variations, the operational variations have a greater effect to the sound power level. As it may be seen, the data can be grouped into three categories in terms of level. The first containing only the measurement in the lowest input voltage, which resulted to the lowest sound power level. In the second category there are the variations for the 50 Hz current frequency, which resulted in similar sound power levels. The third category are the 60 Hz variations, which have more widespread and higher sound power levels than the other categories. By closely examining the FFT spectrum of Figure 4.18, it is apparent that the fan rotation speed changes results also by shifting frequencies from 63 Hz to 250 Hz. The ripples in the high frequencies are attributed to arc reflections. Similarly to the ambient temperature analysis, this frequency segment is plotted against frequency and against frequency over rotation speed (Figure 4.19).

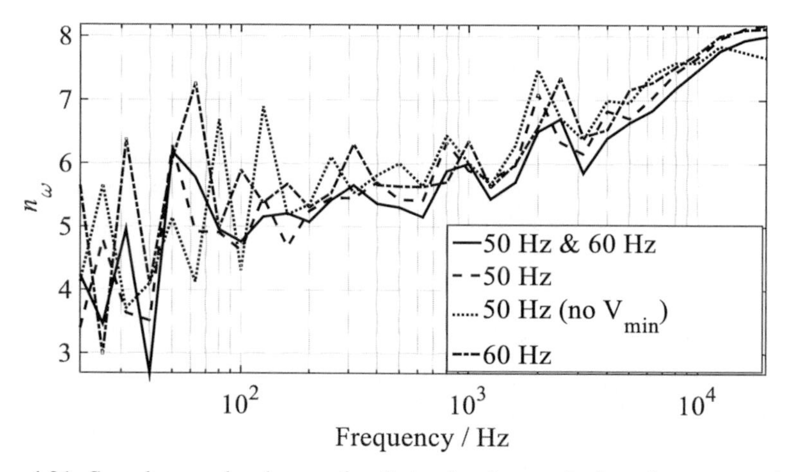

Figure 4.21: Sound power level correction factor for changes in the reference sound source fan rotation speed for four different data sets in one-third octave bands.

An analysis was performed, which shows that the frequency shift is proportional to the rotation speed. Three peaks of the spectrum corresponding to the nominal 50 Hz input voltage (230 V) were chosen (at 81, 163 and 203 Hz). The corresponding peak frequencies were located at the other spectra and the sound power level at the nominal voltage was subtracted by the rest. Figure 4.20 shows the sound power level difference against the rotation speed ratio. According to Figure 4.20 the sound power level depends on the fan rotation speed, which also affects the frequency location of the studied peaks.

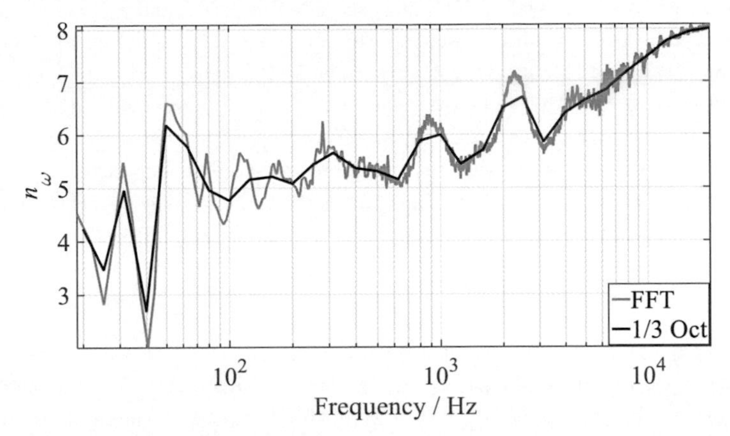

Figure 4.22: Sound power level correction factor for changes in reference sound source fan rotation speed.

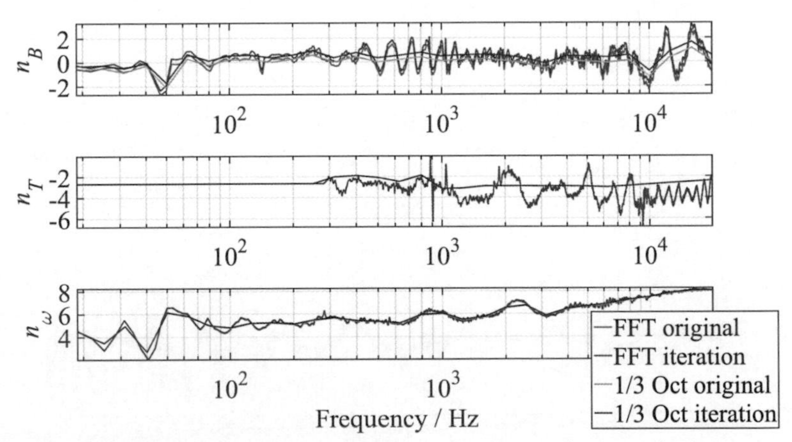

Figure 4.23: Sound power level correction factor for changes in atmospheric pressure (top), ambient temperature (middle) and fan rotation speed (bottom).

Based on the previous, the factor n_ω was calculated similarly to n_T . The frequency regions from 20 Hz to 63 Hz and from 250 Hz to 20 kHz were used for the regression against frequency, since no frequency shift is observed by changing rotation speed. For the interval between 63 Hz and 250 Hz the analysis was performed for the f/ω vector. The frequency shift includes only low frequencies, where the one-third octave bands could not be used due to the small number of samples. For this reason, the factor

n_ω was initially determined in FFT bands and afterwards in one-third octave bands. The correction factor has been calculated for four data sets including: a) all input voltages, b) only the 50 Hz input voltages, c) the 50 Hz input voltages excluding the lowest and d) only the 60 Hz input voltages. The values are presented in Figure 4.21 for one-third octave band analysis. The values of n_ω differ among data sets but in overall the correction factor increases in similar way as frequency increases. In the author's opinion the most descriptive analysis is the one including all measured data and this is to be considered as the proposed one after the fan rotation speed variations. The proposed correction factor is shown in Figure 4.22 in both broad band and narrowband frequency analysis.

The observation of Figure 4.22 reveals that the investigated RSS has varying radiation characteristics in terms of frequency. At frequencies below 100 Hz the correction factor exhibits large variations. From 100 Hz to 1600 Hz it takes values close to 5.5, which correspond to a mixed monopole-dipole behaviour [Hü981] and above 2 kHz the values increase meaning an increase of the radiation order too. The non-unique radiation order comes in tandem with the temperature variations investigation.

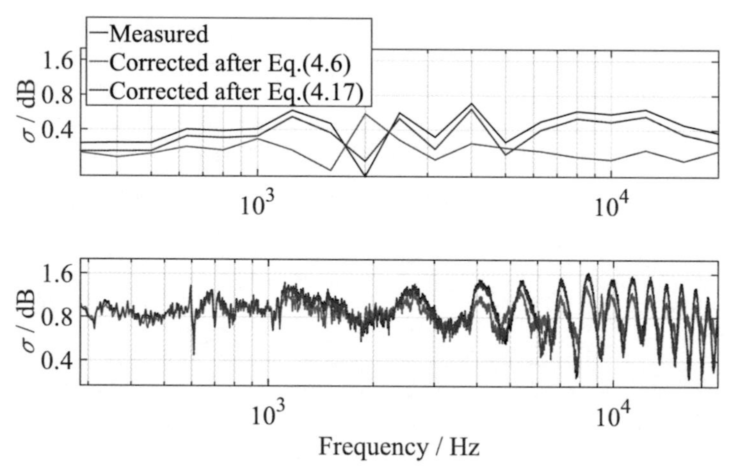

Figure 4.24: Standard deviation of measured and corrected data after Eqs.(4.6) and (4.17) for measurements at various ambient temperatures. B&K source.

4.3.6 Comparison to existing correction

The calculation of the correction factors of Eq.(4.6) included the use of theoretical values for the factors not of interest (e.g. the ambient temperature and fan rotation speed for the calculation of the factor related to atmospheric pressure). For the validation of the results, the factors were calculated once more by replacing the theoretical values with the calculated ones (in the sense of iteration). The results are shown in Figure 4.23. The comparison of the factors shows no significant differences for the ambient temperature and fan rotation speed variations. On the contrary, the factor related to atmospheric pressure variations differs between the calculations and this is attributed to the small variety of atmospheric pressure values at which the sound power measurements were performed.

The proposed correction has been compared to an existing one [Brü13], which is described by:

$$L_{W,\,\text{in situ}} = L_{W,\,\text{cal}} + 10\lg\left(\frac{B_{\text{in situ}}}{B_{\text{cal}}}\right)\text{dB} + 5\lg\left(\frac{T_{\text{cal}}}{T_{\text{in situ}}}\right)\text{dB} + 0.5\left(\omega_{\text{in situ}} - \omega_{\text{cal}}\right)\text{dB} \quad (4.17)$$

where $B_{\text{cal}} = 102.3$ kPa, $T_{\text{cal}} = 20°$ C and $\omega_{\text{cal}} = 46.9$ Hz as given in [Brü13].

The measured data was the in situ values for the Eqs.(4.6) and (4.17). After the calculation of the sound power levels under calibration conditions the comparison of the two applied corrections was performed by calculating the standard deviation of the three sound power level sets (measured, corrected after Eq.(4.6) and corrected after Eq.(4.17)). Figure 4.24 shows the comparison for the ambient temperature measurements and Figure 4.25 for the fan rotation speed measurements both for the Brüel & Kjær source. The standard deviation comparison for the temperature measurements for the other two RSSs is shown in Figure C.10 and Figure C.11 in Appendix C.

The comparison of the two corrections reveals that the one according to Eq.(4.6) provides close sound power levels referred to calibration conditions. This is enhanced by Figure 4.25, where the proposed correction by this study is related to lower standard deviation values especially for frequencies above 3 kHz.

4.4 Near field effects

The present study aims, among others, at the extension of the valid frequency limit of sound power measurements below the present limits (100 Hz) to frequencies lower than 50 Hz. In this frequency region, near field effects pose an important factor affecting the substitution method. A study of such effects was performed using PTB's scanning apparatus. Three RSSs were used (a Brüel & Kjær, the EDF and the Norsonic) and sound pressure measurements were performed at the following radii: 0.60, 0.70, 0.80, 0.90, 1.00, 1.45, 1.70, 2.00, 2.43 & 2.75 m for both one-third octave band and FFT analysis. Windscreens were placed on the microphones (60 mm diameter). The sound pressure values were corrected for air absorption [ISO3745] and the data for the two farthest radii (2.43 & 2.75 m) required two additionally corrections. Firstly, an off-axis correction because the diaphragm was pointing the opposite direction compared to the microphones for the other measurement radii (factor $C_{\mathrm{ang},i}$ in Eq.(3.5)), which was calculated according to the frequency response of the microphones [Brü822]. Secondly, a correction for the attenuation imposed by the arc (factor C_{att} in Eq.(3.5)). This correction was experimentally calculated by measurements with and without the arc. For the first measurements, the arc of Figure 4.7 was used in PTB's hemianechoic room. For the measurements without the arc, microphone stands were used instead. The correction in one-third octave bands was the calculated by:

$$C_{\mathrm{att}} = \overline{L_{p,\,\mathrm{arc}} - L_{p,\,\mathrm{stand}}} \tag{4.18}$$

The FFT values were defined by linear interpolation of the one-third octave band values. Since both correction effects are influential at high frequencies, the values below 1 kHz were set to zero.

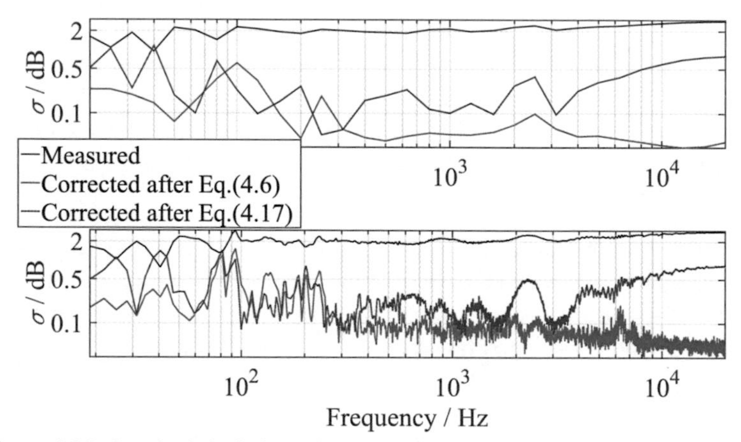

Figure 4.25: Standard deviation of measured and corrected data after Eqs.(4.6) and (4.17) for measurements at various fan rotation speeds. B&K source.

In case of a concentric sphere measured at distance r, the near field error according to Hübner [Hü731] is generally described by:

$$\Delta(r) = 10\lg\left(\frac{2}{\pi k r \left| Ha_{n_{\mathrm{scr}}+\frac{1}{2}}^{(2)}(kr)\right|^2}\right) \mathrm{dB} \tag{4.19}$$

where $Ha_{n_{\mathrm{scr}}+\frac{1}{2}}^{(2)}$ is the normalized Bessel function of the third kind (also known as Hankel function) and n_{scr} is the order of the source (0 for monopole, 1 for dipole etc.).

The near field effects can be quantified by the difference between the determined sound power at each measurement radius and the sound power at the shortest radius according to:

$$\Delta L_{W,\,\mathrm{meas},\,i} = L_W(r_i) - L_W(r_{\min}) \tag{4.20}$$

Eq.(4.19) may be the accordingly used:

$$\Delta L_{W,\,\mathrm{theo},\,i} = \Delta(r_{\min}) - \Delta(r_i) \tag{4.21}$$

Figure 4.26 shows the sound power level difference for both measured and theoretical data for the 1.45 m measurement radius. The difference comparison for all measurement radii reveals the same trend. As it is shown, the sound power level difference as calculated by the measured data is related to a dipole behaviour. It is also seen that below $kr = 10^{0.4}$ the theoretical sound power level difference differs from the measured. This is attributed to remaining wind effects, since the windscreens cannot fully suppress the generated by the RSS wind and to room mode effects. The remaining wind effects could be investigated by correlation measurements. Apparently, a near field correction may be provided by Eq.(4.19) down to the *kr* value where no remaining wind and room mode effects are present. For this reason, in the uncertainty calculation of section 3.9.12 the near field effects were taken into consideration as explained in section 3.9.5.

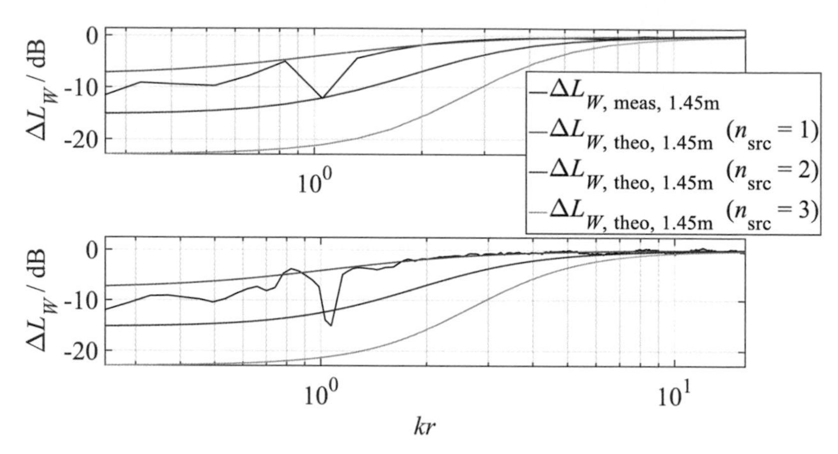

Figure 4.26: Sound power level difference for measured and theoretical data for different source orders against kr. Top: one-third octave bands. Bottom: FFT bands.

4.5 Uncertainty of sound power levels emitted by transfer standards

The uncertainty of the sound power level of transfer standards may be divided into the uncertainty related to the number of microphones used for the determination of the surface and time averaged sound pressure level over the measurement surface and the uncertainty for the calculation of the in situ sound power level based on sound power levels under calibration conditions. Each of these parameters is calculated and presented in the following paragraphs.

4.5.1 Uncertainty due to directivity

The directivity characteristics of a source contribute to the uncertainty for the surface and time averaged sound pressure level. As an example, a directional source requires an adequate measurement sampling (number of microphones) so that the directivity is adequately detected. The sound pressure levels recorded by each microphone were used for the related calculation. For the calculation of the surface and time averaged sound pressure level a different number of microphones was used and the uncertainty was calculated by [JCG08]:

$$u^2\left(\text{directivity}\right)=\frac{\sigma^2\left(L_{p,i}\right)}{n_{\text{mic}}} \tag{4.22}$$

where $\sigma^2\left(L_{p,i}\right)$ is the standard deviation of the sound pressure levels for the microphones used and n_{mic} is the total number of used microphones.

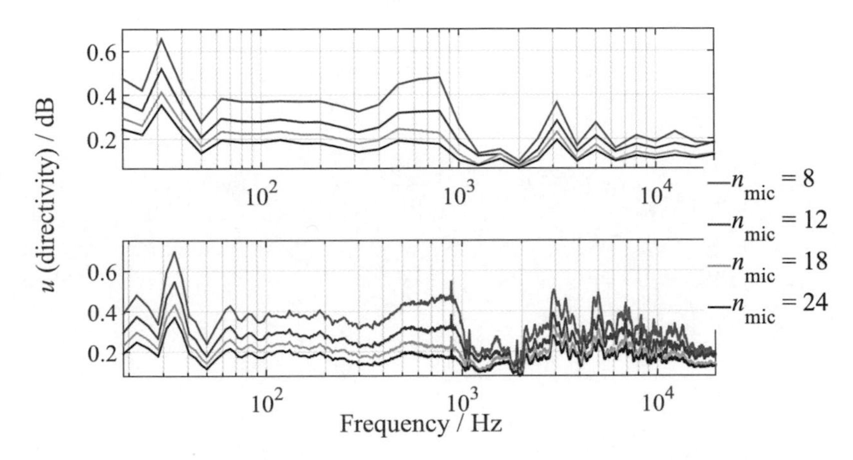

Figure 4.27: Uncertainty of the sound pressure level for different number of microphones for the measurement of the EDF reference sound source. Top: one-third octave bands. Bottom: FFT bands.

The number of microphones used for the surface and time averaged sound pressure level calculation was: 8, 10, 12, 16, 18, 20 & 24. The positions of the microphones are contained in Table C.4. Figure 4.27 shows the uncertainty for the EDF RSS.

The use of a larger number of microphones for the sampling of the measurement surface decreases the related uncertainty for both one-third octave band and FFT analysis. The 24 microphones used for the measurements sufficiently cover the directivity of the sources. The decrease in uncertainty is related to the surface and time averaged sound pressure level difference between the different number of microphones setups. Table C.5 shows the range where the sound pressure level difference lies for each source and frequency analysis.

The range decreases while the number of microphones increases for both frequency analyses. Apparently, the use of a larger number of microphones for the sampling of the measurement surface decreases the uncertainty related to the directivity of the source under measurement.

4.5.2 Uncertainty due to in situ sound power determination

According to GUM [JCG08] the uncertainty of the in-situ sound power level of the transfer source according to Eq.(4.6) is described by:

$$\begin{aligned} u^2\left(L_{W,\mathrm{TS,\,in\,situ}}\right) &= u^2\left(L_{W,\mathrm{TS,\,cal}}\right)+\left[10\lg\left(\frac{B_{\mathrm{in\,situ}}}{B_{\mathrm{cal}}}\right)\mathrm{dB}\right]^2 u^2\left(n_B\right) \\ &+\left[\frac{10\,n_B}{\ln 10\,B_{\mathrm{in\,situ}}}\mathrm{dB}\right]^2 u^2\left(B_{\mathrm{in\,situ}}\right)+\left[\frac{10\,n_B}{\ln 10\,B_{\mathrm{cal}}}\mathrm{dB}\right]^2 u^2\left(B_{\mathrm{cal}}\right) \\ &+\left[10\lg\left(\frac{T_{\mathrm{in\,situ}}}{T_{\mathrm{cal}}}\right)\mathrm{dB}\right]^2 u^2\left(n_T\right)+\left[\frac{10\,n_T}{\ln 10\,T_{\mathrm{in\,situ}}}\mathrm{dB}\right]^2 u^2\left(T_{\mathrm{in\,situ}}\right) \\ &+\left[\frac{10\,n_T}{\ln 10\,T_{\mathrm{cal}}}\mathrm{dB}\right]^2 u^2\left(T_{\mathrm{cal}}\right)+\left[10\lg\left(\frac{\omega_{\mathrm{in\,situ}}}{\omega_{\mathrm{cal}}}\right)\mathrm{dB}\right]^2 u^2\left(n_\omega\right) \\ &+\left[\frac{10\,n_\omega}{\ln 10\,\omega_{\mathrm{in\,situ}}}\mathrm{dB}\right]^2 u^2\left(\omega_{\mathrm{in\,situ}}\right)+\left[\frac{10\,n_\omega}{\ln 10\,\omega_{\mathrm{cal}}}\mathrm{dB}\right]^2 u^2\left(\omega_{\mathrm{cal}}\right) \end{aligned} \tag{4.23}$$

The analysis follows the initial work done by Wittstock [Wit041]. The uncertainty of the sound power level under calibration conditions has been already discussed and presented in paragraph 3.9.12. The uncertainty of the correction factors was calculated based on the error of the regression analysis. This relates the sound power level difference of the measured data and sound power level difference after the regression. As an example, the uncertainty for the correction factor for the influence of atmospheric pressure is given:

$$u^2\left(n_B\right)=\sigma^2\left[\Delta L_W\left(B\right)-\Delta L_{W,\,\mathrm{regression}}\left(B\right)\right] \tag{4.24}$$

where $\Delta L_W\left(B\right)$ is the sound power level difference described by Eq.(4.7) and $\Delta L_{W,\,\mathrm{regression}}\left(B\right)$ the sound power level difference after applying the regression analysis.

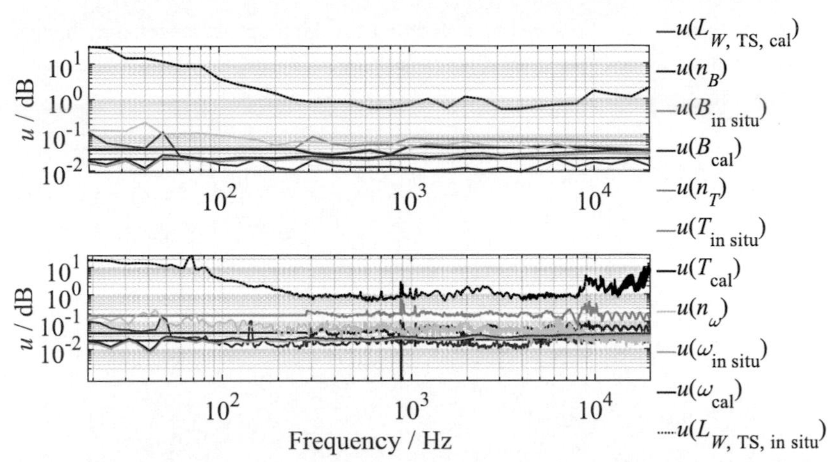

Figure 4.28: Combined uncertainty and its components for the in situ sound power level correction. Top: one-third octave bands. Bottom: FFT bands.

As uncertainty of the correction factor for the variations in atmospheric pressure and ambient temperature, the maximum value of the corresponding individual uncertainties (one uncertainty pro RSS) was chosen. The atmospheric pressure, ambient temperature and fan rotation speed were taken from Table C.1, Table C.2 and Table C.3. As in situ values the ones corresponding to the related parameter under investigation were chosen, leaving the other two sets as values under calibration conditions. E.g. for the variations in atmospheric pressure the atmospheric pressure values of Table C.1 were chosen as in situ values, while the values of Table C.2 and Table C.3 as values under calibration conditions. From each in situ value set the maximum value was chosen and from each set with calibration condition values the minimum, so as to maximize the related ratio of Eq.(4.23). The uncertainties of the atmospheric and operational parameters are related to the uncertainty of the measurement instrument. These are: $u(B) = 0.3$ kPa, $u(T) = 0.5$ K and $u(\omega) = 0.05$ Hz. Figure 4.28 shows the uncertainty of the in situ sound power level as described by Eq.(4.23). As it may be observed, in broadband analysis the most influential factor is the uncertainty of the sound power level under calibration conditions. In narrow band analysis, the uncertainty of the ambient temperature correction factor also affects the overall in situ uncertainty but only around 1 kHz and to a lesser extent compared to the uncertainty of the sound power level under calibration conditions.

The uncertainty of the correction factor for the variation in fan rotation speed was calculated for each of the four fan ration speed measurement groups as discussed in section 4.3.5 and the results are presented in Figure 4.29. The differences among the uncertainties are small and it can be stated that for the proposed uncertainty budget there is no need for separate uncertainty calculation referring to different alternating frequencies.

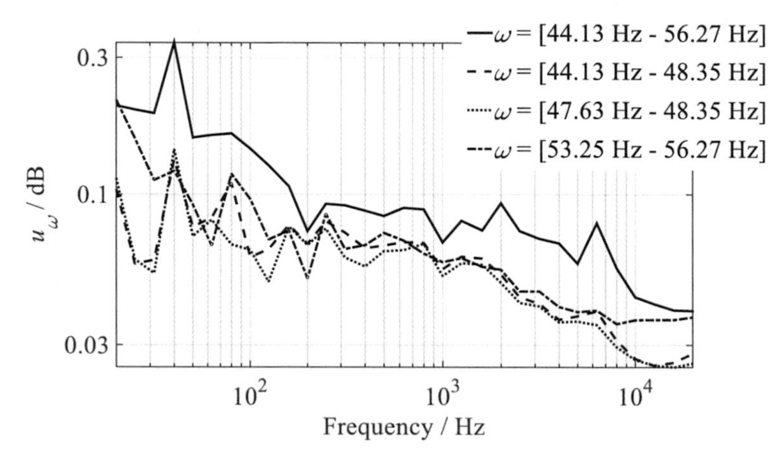

Figure 4.29: Uncertainty of the correction factor for changes in fan rotation speed for various rotation speed values.

The determination of the sound power level of the transfer standard in situ enables the determination of sound power of real sources. The sound power determination of equipment and machinery by applying the substitution method using both sound pressure and sound intensity are the main topic the following chapter. The influence of the measurement surface and the surrounding environment was investigated for a number of sources with various spectral contents. The determination of the related uncertainty enabled the combined uncertainty for the dissemination of the unit watt to be derived. The combined uncertainty was compared to the existing uncertainty values and the possibility of a new uncertainty budget for the sound power determination is discussed.

5 Implementation of substitution method

The substitution method was applied using both sound pressure and sound intensity measurements and the results are compared in this chapter. The influence of the measurement surface and the surrounding environment is discussed. The related uncertainties are also presented.

5.1 Influence of the measurement surface

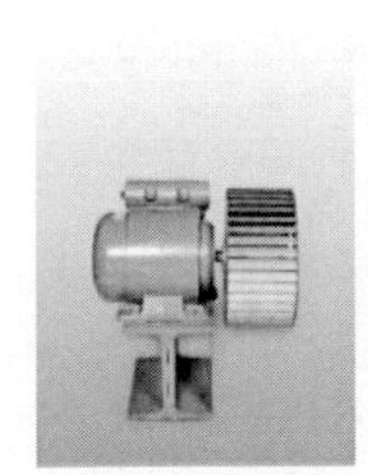
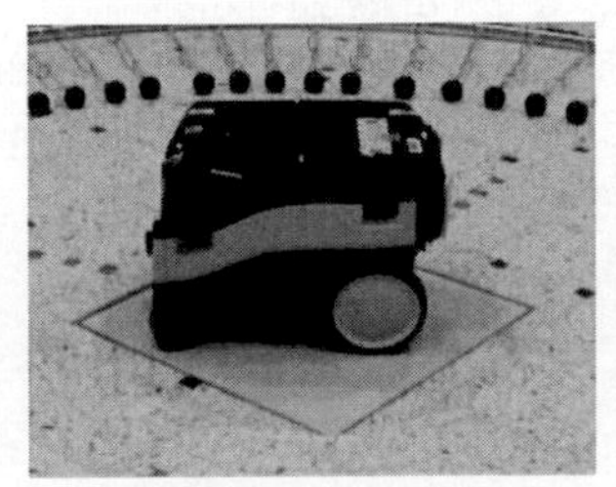

Figure 5.1: Realistic sources used for the application of the substitution method. Left to right: aerodynamic reference sound source, vacuum cleaner and air compressor.

Table 5.1 Measurement points coordinates for the box shaped measurement surface.

x (m)	y (m)	z (m)
-0.625	-1.25	0.4
0.625	-1.25	1.1
1.25	-0.625	1
1.25	0.625	0.5
0.625	1.25	0.6
-0.625	1.25	0.9
-1.25	0.625	0.8
-1.25	-0.625	0.7
-0.625	0.625	1.5
-0.625	-0.625	1.5
0.625	0.625	1.5
0.625	-0.625	1.5

Five sources were used for the implementation of the substitution method using both sound pressure and sound intensity measurements. PTB's primary source [Kir06], two RSSs (an AEG Type ADEB71K2 and a Brüel & Kjær Type 4204) as shown in Figure 3.1, a vacuum cleaner (Festool Type CMT 22E) and an air compressor (Kaeser Type Diamant 160W). The two latter sources were chosen for their tonal spectral components. The Brüel & Kjær source was used as the transfer standard for the determination of the in situ sound power level of the realistic sources (AEG, vacuum cleaner and air compressor) based on the free field sound power level of the primary source. Figure 5.1 shows the realistic sources.

The sound intensity calculation is mentioned in paragraph 3.2.2. The sound power determination based on sound intensity measurements is described in the ISO 9614 series ([ISO96141], [ISO96142], [ISO96143]). The measurement may be performed either by scanning or at discrete points of a surrounding surface. Sound intensity measurements were performed in PTB's hemianechoic room using the scanning apparatus along with measurements at discrete points. The former covered a hemispherical surface, whereas the latter a twelve-point box shaped surface with dimensions 2.5 m x 2.5 m x 1.5 m. Table 5.1 contains the measurement points coordinates for the box shaped measurements.

The sound intensity level for both occasions is given by:

$$L_I = 10\log\left(\frac{|I_\mathrm{r}|}{I_\mathrm{ref}}\right)\mathrm{dB} \tag{5.1}$$

where $|I_\mathrm{r}|$ is the unsigned value of the radial (normal) component of the sound intensity [ISO96143].

The sound power determination of the realistic sources (or devices under test, DUT) includes two steps. In the first, the sound power level of the transfer source under calibration conditions is determined. For the case of sound intensity measurements, the sound power level is given by:

$$\begin{aligned} L_{W,\,\mathrm{TS,\,cal},\,I} &= L_{W,\,\mathrm{PS}} + \overline{L_{I,\,\mathrm{TS,\,cal}}} - \overline{L_{I,\,\mathrm{PS}}} + 10\lg\left(\frac{S_\mathrm{TS,\,cal}}{S_\mathrm{PS}}\right)\mathrm{dB} \\ &-10\lg\left(\frac{B_\mathrm{TS,\,cal}}{B_\mathrm{PS}}\right)\mathrm{dB} + 5\lg\left(\frac{T_\mathrm{TS,\,cal}}{T_\mathrm{PS}}\right)\mathrm{dB} + C_{\mathrm{scr,\,probe},\,I} + C_{3,\,\mathrm{TS,\,cal}} - C_{3,\,\mathrm{PS}} + C_\mathrm{FFT} \end{aligned} \tag{5.2}$$

whereas for sound pressure measurements:

$$\begin{aligned} L_{W,\,\mathrm{TS,\,cal},\,p} &= L_{W,\,\mathrm{PS}} + \overline{L_{p,\,\mathrm{TS,\,cal}}} - \overline{L_{p,\,\mathrm{PS}}} + 10\lg\left(\frac{S_\mathrm{TS,\,cal}}{S_\mathrm{PS}}\right)\mathrm{dB} \\ &-10\lg\left(\frac{B_\mathrm{TS,\,cal}}{B_\mathrm{PS}}\right)\mathrm{dB} + 5\lg\left(\frac{T_\mathrm{TS,\,cal}}{T_\mathrm{PS}}\right)\mathrm{dB} + C_\mathrm{noise,\,TS} - C_\mathrm{noise,\,PS} \\ &+ C_{\mathrm{scr,\,probe},\,p} + C_{3,\,\mathrm{TS,\,cal}} - C_{3,\,\mathrm{PS}} + C_\mathrm{FFT} \end{aligned} \tag{5.3}$$

In the second step, the sound power level of the DUT is determined, in case of sound intensity, by:

$$\begin{aligned}L_{W,\text{DUT, cal},I} &= L_{W,\text{TS, cal},I} + \overline{L_{I,\text{DUT, cal}}} - \overline{L_{I,\text{TS, cal}}} + 10\lg\left(\frac{S_{\text{DUT, cal}}}{S_{\text{TS, cal}}}\right)\text{dB} \\ &-10\lg\left(\frac{B_{\text{DUT, cal}}}{B_{\text{TS, cal}}}\right)\text{dB} + 25\lg\left(\frac{T_{\text{DUT, cal}}}{T_{\text{TS, cal}}}\right)\text{dB} + C_{3,\text{DUT, cal}} - C_{3,\text{TS, cal}}\end{aligned} \tag{5.4}$$

and in case of sound pressure by:

$$\begin{aligned}L_{W,\text{DUT, cal},p} &= L_{W,\text{TS, cal},p} + \overline{L_{p,\text{DUT, cal}}} - \overline{L_{p,\text{TS, cal}}} \\ &+10\lg\left(\frac{S_{\text{DUT, cal}}}{S_{\text{TS, cal}}}\right)\text{dB} - 10\lg\left(\frac{B_{\text{DUT, cal}}}{B_{\text{TS, cal}}}\right)\text{dB} + 25\lg\left(\frac{T_{\text{DUT, cal}}}{T_{\text{TS, cal}}}\right)\text{dB} \\ &+C_{\text{noise, DUT, cal}} - C_{\text{noise, TS, cal}} + C_{3,\text{DUT, cal}} - C_{3,\text{TS, cal}}\end{aligned} \tag{5.5}$$

where S_x is the surface over which the sound power of the subscripted source is determined respectively and $C_{\text{scr, probe},I}$ and $C_{\text{scr, probe},p}$ is the intensity probe windscreen correction factor for sound intensity and sound pressure respectively. It is similar to the correction discussed in section 3.9.6 but of different value, because of the different windscreen shape (the pressure microphone windscreens are spherical, while the sound intensity probe are oval with two axes of symmetry). The near field effects and the remaining room influences have not been included as a correction factor but are taken into consideration in the related uncertainty instead.

It must be noted that as sound pressure value, the mean value of the intensity probe microphone signals was used. This way, the arc attenuation as discussed in paragraph 4.4 is compensated, due to the different distance of each intensity probe microphone from the arc.

As it may be observed, Eqs.(5.2)-(5.5) include corrections for the atmospheric pressure and ambient temperature. Eqs.(5.2) and (5.3) express the transfer standard sound power level under calibration conditions, with the ambient temperature correction exponent to be -0.5 due to the different order of the sources (primary source a monopole and transfer source not a monopole). Eqs.(5.4) and (5.5) provide the sound power level of the DUT under calibration conditions. After spherical harmonics decomposition, the order of the vacuum cleaner and the air compressor was found to be similar to the transfer source and the temperature correction factor was set to -2.5 (dipole).

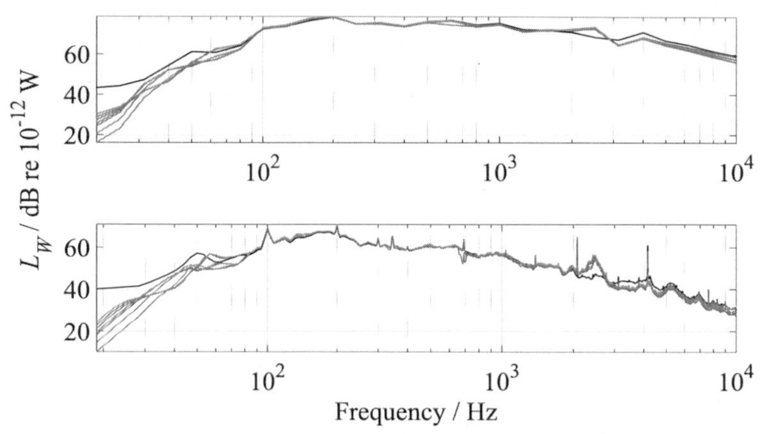

Figure 5.2: Sound power level of the vacuum cleaner directly determined after sound pressure scanning measurements (black) and after applying the substitution method (grey). Top: one-third octave bands. Bottom: FFT bands.

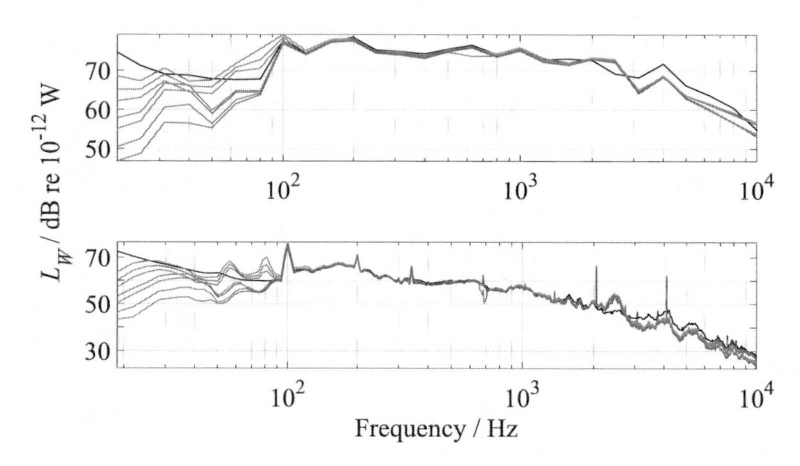

Figure 5.3: Sound power level of the vacuum cleaner directly determined after sound intensity scanning measurements (black) and after applying the substitution method (grey). Top: one-third octave bands. Bottom: FFT bands.

The primary source was measured only by scanning at 0.70 m, 0.80 m, 0.90 m, 1.45 m, 1.70 m, 2.34 m & 2.75 m measurement radius. The transfer source was measured by scanning at the same radii and with the box shaped surface, while the DUT by scanning only at 1.45 m and with the rectangular box. The implementation of Eqs.(5.2) and (5.3) could include all possible combinations between measurements at different radii. This was chosen not to be the case for two reasons. First, because this would not be representative for realistically applicable engineering measurements, meaning that for time convenience, only a radius would be chosen for all measurements. Second, because the large number of the available sound pressure or sound intensity level differences would have an impact by decreasing the related uncertainty.

The sound power level of the primary source was determined by measuring the piston vibration velocity using a laser scanning vibrometer. The measurements were not part of this project and the sound power values were kindly provided for the implementation of the substitution method.

The sound power levels after the substitution method were compared to the directly calculated using Eq.(4.3) for the case of sound pressure and the following equation for the case of sound intensity [ISO96143]:

$$L_W = 10\lg\left(\frac{\left|\sum_{i}^{N} \overline{I_{\mathrm{r}i}}\, S_i\right|}{P_{\mathrm{ref}}}\right)\mathrm{dB} \tag{5.6}$$

where $I_{\mathrm{r}i}$ is the signed magnitude of the partial surface average radial sound intensity measured on the partial surface i of the measurement surface and S_i is the area of the partial surface i.

Table D.1 shows the atmospheric pressure, ambient temperature and surface values used for the corrections in Eqs.(5.2)-(5.5).

Figure 5.2 shows the sound power level of the vacuum cleaner after direct determination and after the substitution method for the scanning measurements using sound pressure measurements and Figure 5.3 using sound intensity measurements. Figure 5.4 and Figure 5.5 show the same levels for the partial surface measurements. The corresponding figures for the other two DUT can be found in Figure D.1-Figure D.8 in Appendix D. As it can be seen, the directly determined sound power level is close to the levels after the substitution method in case of the scanning measurements for the frequency region between 100 Hz and 1600 Hz, but not for the case of sound pressure measurements over the box shaped surface. The sound intensity results are close for frequencies below 100 Hz as well. The deviation between the direct and after the substitution sound power levels at high frequencies seen at both measurement sets are attributed to the inefficient frequency response of the primary sound source. Room influences may be seen at frequencies below 100 Hz for the sound pressure measurements, with the results based on the box shaped surface being more influenced than those after scanning. The latter smoothens the room effects. Near field effects also affect the results along with remaining wind influences carried mainly by the transfer standard data. The substitution method results are not largely deviated in case of scanning, especially at frequencies above 100 Hz. The sound intensity results of the box shaped surface are also not much deviated. It may also be observed that the narrow band frequency analysis provides qualitative results.

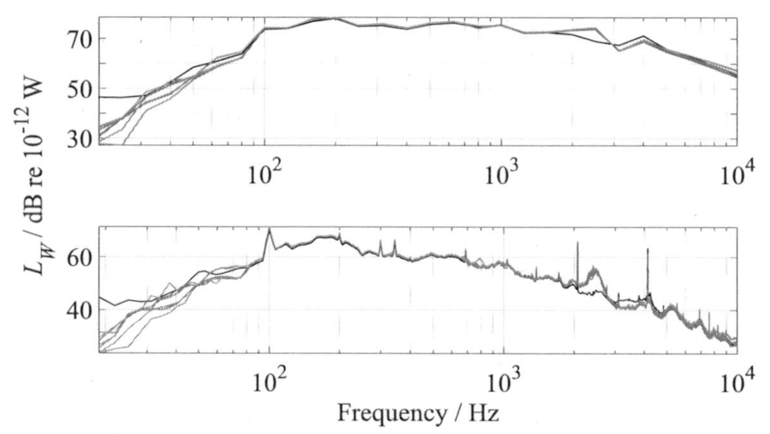

Figure 5.4: Sound power level of the vacuum cleaner directly determined after sound pressure discrete point measurements (black) and after applying the substitution method (grey). Top: one-third octave bands. Bottom: FFT bands.

5.2 Influence of the surrounding environment

Apart from influences of the measurement surface, the influences of the surrounding environment were also studied. The same sources were used as previously, and the measurements were performed in various environments of different volume and absorption. The spaces where the measurements took place are described in Table 5.2.

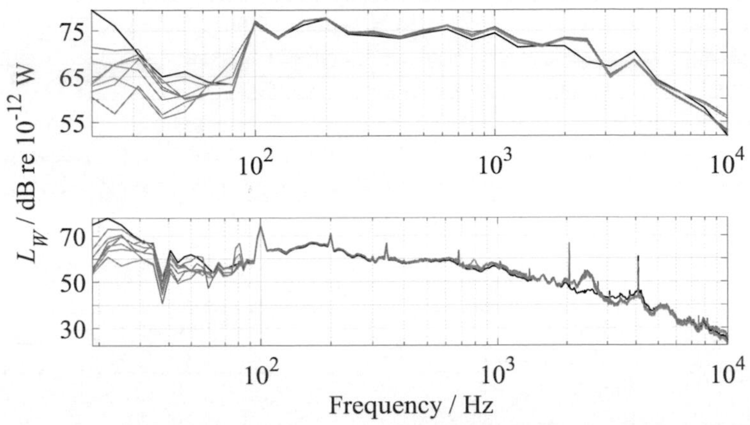

Figure 5.5: Sound power level of the vacuum cleaner directly determined after sound intensity discrete point measurements (black) and after applying the substitution method (grey). Top: one-third octave bands. Bottom: FFT bands.

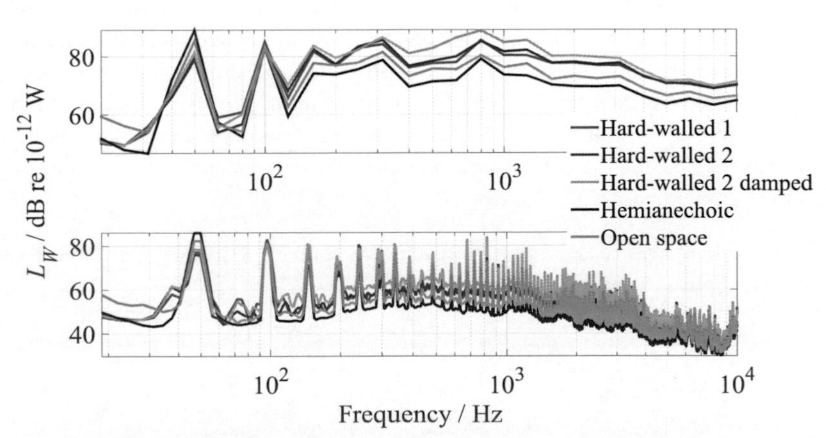

Figure 5.6: Sound power level of the air compressor directly determined after sound pressure discrete point measurements at different acoustic environments. Top: one-third octave bands. Bottom: FFT bands.

The influences of the surrounding environment may be seen in Figure 5.6 and Figure 5.7 as the resulting differences on the directly calculated sound power level of the air compressor based on sound pressure and sound intensity measurements respectively. The corresponding levels for the other two DUT can be found in Figure D.9-Figure D.12 in Appendix D. In the results no environmental correction has been applied (correction factor K_2 as it may be found in the ISO 3740 series ([ISO3744], [ISO3746]). This is because there is no intention to directly compare the proposed sound power level determination method to existing ones and because the determination of K_2 would be cumbersome in case of the open space measurements.

Table 5.2 Description of measurement surroundings.

Room	**Volume (m³)**
Hard-walled test room 1	50
Hard-walled test room 2	50
Hard-walled test room 2 – damped	50
Hemianechoic room	190
Open space in large hall	-

The sound power analysis included the comparison between directly determined sound power levels based on sound pressure and sound intensity measurements and sound power levels after applying the substitution method. For the substitution method, the determination of the sound power level of the transfer source was once performed under calibration conditions and once in situ conditions. For the relation of the two, the correction for the changes in environmental and operational conditions was performed as expressed by Eq.(4.6). Table D.2 shows the environmental and operational conditions and the measurement surface values for the measurements in different environmental surroundings.

The substitution procedure for the determination of the DUT sound power level in situ for sound pressure measurements is described by:

$$L_{W,\text{DUT, in situ},p} = L_{W,\text{TS, in situ}} + \overline{L_{p,\text{DUT, in situ}}} - \overline{L_{p,\text{TS, in situ}}} \\ +10\lg\left(\frac{S_{\text{DUT, in situ}}}{S_{\text{TS, in situ}}}\right)\text{dB} - 10\lg\left(\frac{B_{\text{DUT, in situ}}}{B_{\text{TS, in situ}}}\right)\text{dB} + 25\lg\left(\frac{T_{\text{DUT, in situ}}}{T_{\text{TS, in situ}}}\right)\text{dB} \qquad (5.7)$$

The same equation applies for sound intensity measurements. The identical environmental conditions referring to the transfer source and the DUT along with the same background noise levels, make the corrections for the background noise and the air absorption to vanish.

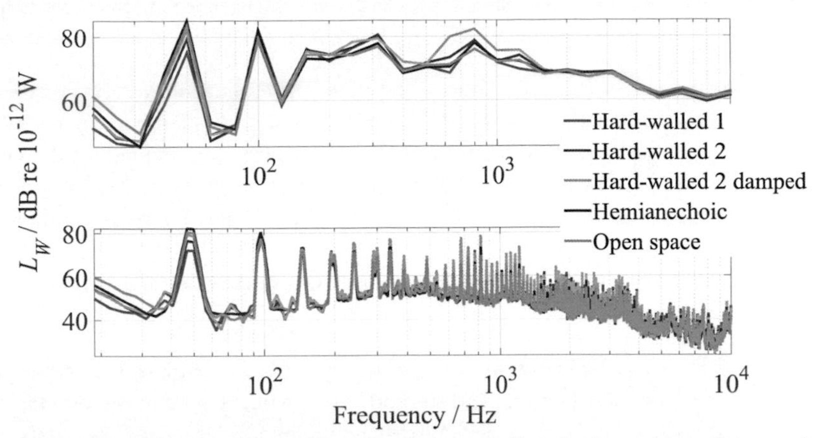

Figure 5.7: Sound power level of the air compressor directly determined after sound intensity discrete point measurements at different acoustic environments. Top: one-third octave bands. Bottom: FFT bands.

Figure 5.8 and Figure 5.9 show the sound power levels of the AEG source after applying both methods. The corresponding results for the other two sources are depicted in Figure D.13-Figure D.16 in Appendix D. The primary source frequency response effects are apparent above 2 kHz for the substitution method results. As expected, the sound power direct results based on sound pressure measurements are strongly influenced by the surrounding environment. The closest to the substitution method results is the hemianechoic room sound power as also discussed in the previous section. The sound power levels based on intensity levels are close to each other except for frequencies below 60 Hz where the sound power levels of the direct calculations are larger than the substitution method results.

5.3 Comparison between sound pressure and sound intensity results

The difference between the sound power levels after the substitution method and after direct determination was calculated as:

$$\Delta L_W = L_{W,\,\text{sub}} - L_{W,\,\text{dir}} \tag{5.8}$$

The mean value of the sound power level differences of all three DUT was calculated and is presented in the following figures. In Figure 5.10 the sound power level differences for the investigation of the influence of the measurement surface is presented. Figure 5.11 and Figure 5.12 present the mean value for the surrounding environment influence after sound pressure and sound intensity measurements respectively. The primary source influence is apparent to all figures above 1.6 kHz.

The measurement surface affects the sound power determination to a small extent. The directly determined sound power level after sound pressure measurements over the box shaped surface has the largest deviation from the corresponding sound power level after the substitution method. The largest deviations between the two sound power determination methods are seen for sound pressure measurements at different surrounding environments. The lowest deviation for this case is seen in the hemianechoic measurements. The sound intensity measurements are less prone to the surrounding environment.

The most important finding of the comparison of the two determination methods is apparent to both measurement surface and surrounding environment results. At frequencies below 60 Hz, the substitution method provides up to approximately 29 dB lower sound power levels compared to the direct determination method. In other words, the sound power that is really emitted to the measured acoustic field is 29 dB larger than the sound power level that would be emitted in case of free field. As it may also be seen, the smallest deviations between substitution and direct method can be found for the case of free or approximately free field (hemianechoic room and open space). At low frequencies there are three factors that assumingly influence the sound power level difference. The room volume at very low frequencies and the room modes and near field at low frequencies.

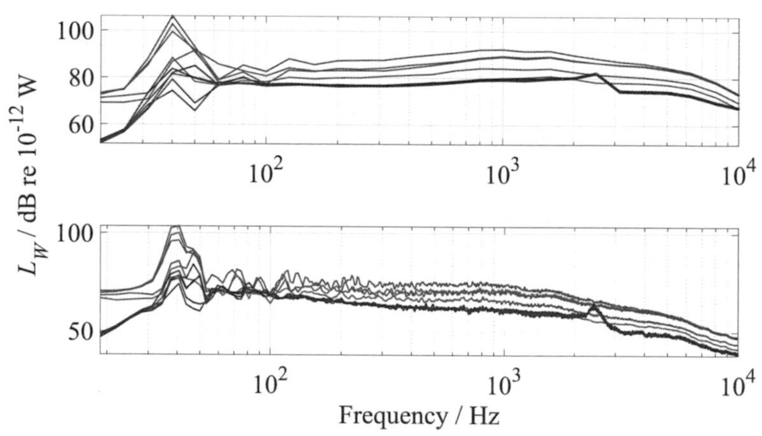

Figure 5.8: Sound power level of the AEG reference sound source directly determined after sound pressure discrete point measurements (blue) and after applying the substitution method (orange red) at different surrounding environments. Top: one-third octave bands. Bottom: FFT bands.

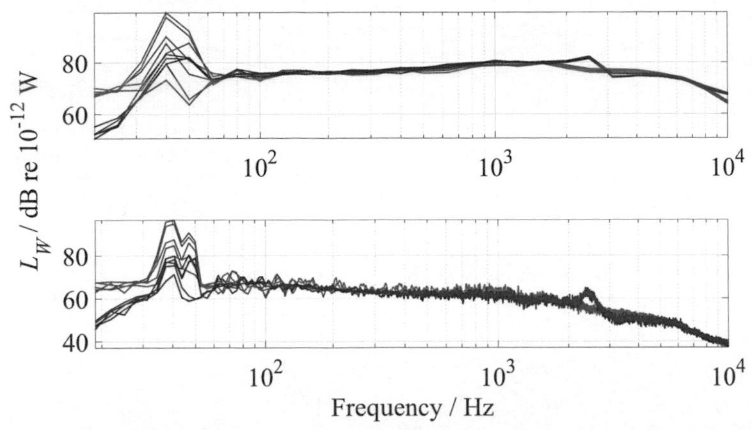

Figure 5.9: Sound power level of the AEG reference sound source directly determined after sound intensity discrete point measurements (blue) and after applying the substitution method (orange red) at different surrounding environments. Top: one-third octave bands. Bottom: FFT bands.

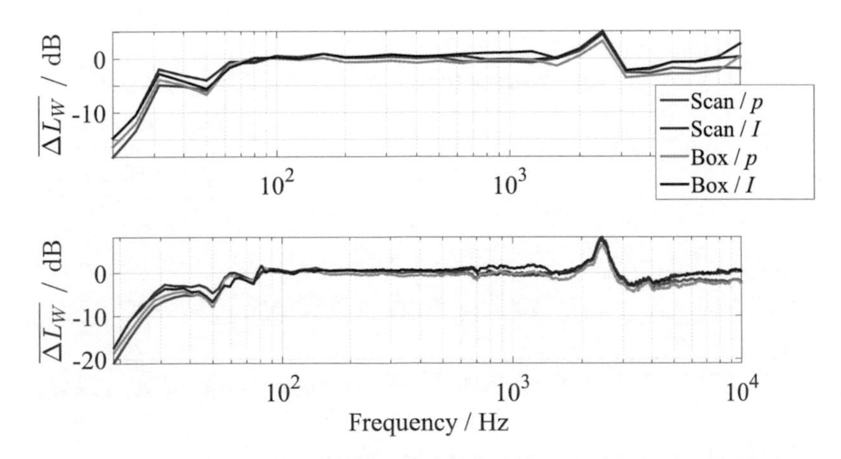

Figure 5.10: Mean sound power level difference between sound power level after the substitution method and after direct calculation for the investigation on the measurement surface influence. Top: one-third octave bands. Bottom: FFT bands.

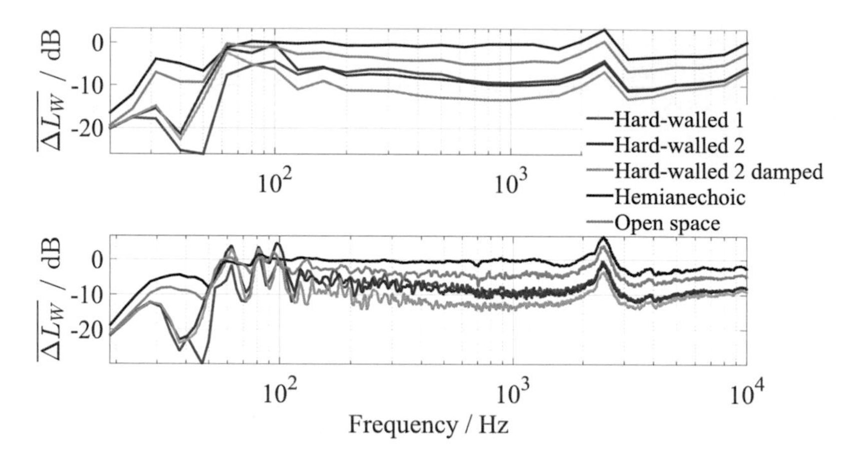

Figure 5.11: Mean sound power level difference between sound power level after the substitution method and after direct calculation for the investigation on the surrounding environment influence for sound pressure measurements. Top: one-third octave bands. Bottom: FFT bands.

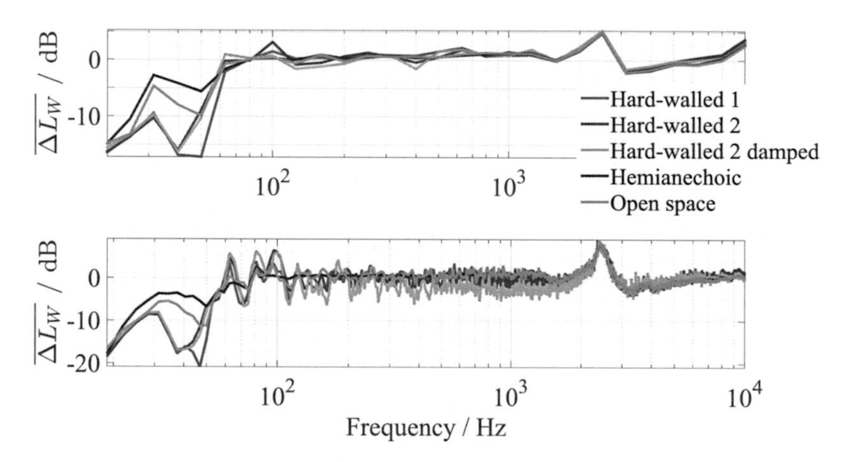

Figure 5.12: Mean sound power level difference between sound power level after the substitution method and after direct calculation for the investigation on the surrounding environment influence for sound intensity measurements. Top: one-third octave bands. Bottom: FFT bands.

5.4 Uncertainty of device under test sound power level

As in section 3.9, the uncertainty of the DUT sound power level in calibration conditions and in situ was performed. In the next section the uncertainty analysis is presented.

5.4.1 Uncertainty of the measurement surface influence

The uncertainty of the transfer source sound power level under calibration conditions as described by Eq.(5.2) is:

$$\begin{aligned}
u^2\left(L_{W,\mathrm{TS,cal},I}\right) &= u^2\left(L_{W,\mathrm{PS}}\right) + u^2\left[\overline{L_{I,\mathrm{TS,cal}}} - \overline{L_{I,\mathrm{PS}}} + 10\lg\left(\frac{S_{\mathrm{TS,cal}}}{S_{\mathrm{PS}}}\right)\mathrm{dB}\right] \\
&+ \left[\frac{10\,\mathrm{dB}}{\ln 10}\frac{B_{\mathrm{PS}}}{B_{\mathrm{TS,cal}}}u^2\left(\frac{B_{\mathrm{TS,cal}}}{B_{\mathrm{PS}}}\right)\right]^2 + \left[\frac{5\,\mathrm{dB}}{\ln 10}\frac{T_{\mathrm{PS}}}{T_{\mathrm{TS,cal}}}u^2\left(\frac{T_{\mathrm{TS,cal}}}{T_{\mathrm{PS}}}\right)\right]^2 \\
&+ u^2\left[\overline{L_{I,\mathrm{TS,cal}}} + 10\lg\left(\frac{S_{\mathrm{TS,cal}}}{S_{\mathrm{TS,cal,ref}}}\right)\mathrm{dB}\right] + u^2\left[\overline{L_{I,\mathrm{PS}}} + 10\lg\left(\frac{S_{\mathrm{PS}}}{S_{\mathrm{PS,ref}}}\right)\mathrm{dB}\right] \\
&+ u^2\left(C_{\mathrm{scr,probe},I}\right) + u^2\left(C_{3,\mathrm{TS,cal}} - C_{3,\mathrm{PS}}\right) + u^2\left(C_{\mathrm{FFT}}\right)
\end{aligned} \tag{5.9}$$

The uncertainty based on sound pressure measurements following Eq.(5.3) is expressed as:

$$\begin{aligned}
u^2\left(L_{W,\mathrm{TS,cal},p}\right) &= u^2\left(L_{W,\mathrm{PS}}\right) + u^2\left[\overline{L_{p,\mathrm{TS,cal}}} - \overline{L_{p,\mathrm{PS}}} + 10\lg\left(\frac{S_{\mathrm{TS,cal}}}{S_{\mathrm{PS}}}\right)\mathrm{dB}\right] \\
&+ \left[\frac{10\,\mathrm{dB}}{\ln 10}\frac{B_{\mathrm{PS}}}{B_{\mathrm{TS,cal}}}u^2\left(\frac{B_{\mathrm{TS,cal}}}{B_{\mathrm{PS}}}\right)\right]^2 + \left[\frac{5\,\mathrm{dB}}{\ln 10}\frac{T_{\mathrm{PS}}}{T_{\mathrm{TS,cal}}}u^2\left(\frac{T_{\mathrm{TS,cal}}}{T_{\mathrm{PS}}}\right)\right]^2 \\
&+ u^2\left[\overline{L'_{p,\mathrm{TS,cal}}} + 10\lg\left(\frac{S_{\mathrm{TS,cal}}}{S_{\mathrm{TS,cal,ref}}}\right)\mathrm{dB}\right] + u^2\left[\overline{L'_{p,\mathrm{PS}}} + 10\lg\left(\frac{S_{\mathrm{PS}}}{S_{\mathrm{PS,ref}}}\right)\mathrm{dB}\right] \\
&+ u^2\left(C_{\mathrm{noise,PS}}\right) + u^2\left(C_{\mathrm{noise,TS}}\right) + u^2\left(C_{\mathrm{scr,probe},p}\right) \\
&+ u^2\left(C_{3,\mathrm{TS,cal}} - C_{3,\mathrm{PS}}\right) + u^2\left(C_{\mathrm{FFT}}\right)
\end{aligned} \tag{5.10}$$

It must be noted that the uncertainty of the sound pressure levels applies to the uncorrected for background noise and air absorption levels. For the calculation of the uncertainty of the sound pressure level difference, Eq.(3.21) was used, where two uncertainty components are included. For the uncertainty of the sound intensity level difference, only the statistical effects were taken into consideration, as expressed by the standard deviation of the differences. To calculate the uncertainty of the sound intensity and sound pressure levels, they must be referred to the same measurement surface. As reference the 1.45 m radius measurement surface was chosen for both primary and transfer sources. The environmental conditions uncertainties were found to be 0.016 dB for the atmospheric pressure and 0.002 dB for the ambient temperature, allowing to be neglected. Thus, the uncertainty equations may be simplified to:

$$\begin{aligned} u^2\left(L_{W,\,\text{TS, cal},\,I}\right) &= u^2\left(L_{W,\,\text{PS}}\right) + u^2\left[\overline{L_{I,\,\text{TS, cal}}} - \overline{L_{I,\,\text{PS}}} + 10\lg\left(\frac{S_{\text{TS, cal}}}{S_{\text{PS}}}\right)\text{dB}\right] \\ &+ u^2\left[\overline{L_{I,\,\text{TS, cal}}} + 10\lg\left(\frac{S_{\text{TS, cal}}}{S_{\text{TS, cal, ref}}}\right)\text{dB}\right] + u^2\left[\overline{L_{I,\,\text{PS}}} + 10\lg\left(\frac{S_{\text{PS}}}{S_{\text{PS, ref}}}\right)\text{dB}\right] \\ &+ u^2\left(C_{\text{scr, probe},\,I}\right) + u^2\left(C_{3,\,\text{TS, cal}} - C_{3,\,\text{PS}}\right) + u^2\left(C_{\text{FFT}}\right) \end{aligned} \tag{5.11}$$

and

$$\begin{aligned} u^2\left(L_{W,\,\text{TS, cal},\,p}\right) &= u^2\left(L_{W,\,\text{PS}}\right) + u^2\left[\overline{L_{p,\,\text{TS, cal}}} - \overline{L_{p,\,\text{PS}}} + 10\lg\left(\frac{S_{\text{TS, cal}}}{S_{\text{PS}}}\right)\text{dB}\right] \\ &+ u^2\left[\overline{L'_{p,\,\text{TS, cal}}} + 10\lg\left(\frac{S_{\text{TS, cal}}}{S_{\text{TS, cal, ref}}}\right)\text{dB}\right] + u^2\left[\overline{L'_{p,\,\text{PS}}} + 10\lg\left(\frac{S_{\text{PS}}}{S_{\text{PS, ref}}}\right)\text{dB}\right] \\ &+ u^2\left(C_{\text{noise, PS}}\right) + u^2\left(C_{\text{noise, TS}}\right) + u^2\left(C_{\text{scr, probe},\,p}\right) \\ &+ u^2\left(C_{3,\,\text{TS, cal}} - C_{3,\,\text{PS}}\right) + u^2\left(C_{\text{FFT}}\right) \end{aligned} \tag{5.12}$$

The uncertainty of the primary source was described in section 3.9.8. The uncertainty of the average sound pressure and sound intensity level is given by Eq.(3.16). The background noise level uncertainty is expressed by Eq.(3.15), the windscreen uncertainties by Eq.(3.19), the air absorption correction uncertainty by Eq.(3.23) and the uncertainty for the FFT windowing by Eq.(3.17).

Similarly, the uncertainty of the DUT sound power level based on sound intensity measurements under calibration conditions using the scanning method is calculated by:

$$
\begin{aligned}
u^2\left(L_{W,\,\mathrm{DUT,\,cal},\,I}\right)_{\mathrm{scan}} &= u^2\left(L_{W,\,\mathrm{TS,\,cal},\,I}\right) \\
&+u^2\left[\overline{L_{I,\,\mathrm{DUT,\,cal}}}-\overline{L_{I,\,\mathrm{TS,\,cal}}}+10\lg\left(\frac{S_{\mathrm{DUT,\,cal}}}{S_{\mathrm{TS,\,cal}}}\right)\mathrm{dB}\right]_{\mathrm{scan}} \\
&+u^2\left(\overline{L_{I,\,\mathrm{DUT,\,cal}}}\right)_{\mathrm{scan}}+u^2\left[\overline{L_{I,\,\mathrm{TS,\,cal}}}+10\lg\left(\frac{S_{\mathrm{TS,\,cal}}}{S_{\mathrm{TS,\,cal,\,ref}}}\right)\mathrm{dB}\right]_{\mathrm{scan}} \\
&+u^2\left(C_{3,\,\mathrm{DUT,\,cal}}-C_{3,\,\mathrm{TS,\,cal}}\right)
\end{aligned}
\tag{5.13}
$$

The related uncertainty for sound pressure is:

$$
\begin{aligned}
u^2\left(L_{W,\,\mathrm{DUT,\,cal},\,p}\right)_{\mathrm{scan}} &= u^2\left(L_{W,\,\mathrm{TS,\,cal},\,p}\right) \\
&+u^2\left[\overline{L_{p,\,\mathrm{DUT,\,cal}}}-\overline{L_{p,\,\mathrm{TS,\,cal}}}+10\lg\left(\frac{S_{\mathrm{DUT,\,cal}}}{S_{\mathrm{TS,\,cal}}}\right)\mathrm{dB}\right]_{\mathrm{scan}} \\
&+u^2\left(\overline{L'_{p,\,\mathrm{DUT,\,cal}}}\right)_{\mathrm{scan}}+u^2\left[\overline{L'_{p,\,\mathrm{TS,\,cal}}}+10\lg\left(\frac{S_{\mathrm{TS,\,cal}}}{S_{\mathrm{TS,\,cal,\,ref}}}\right)\mathrm{dB}\right]_{\mathrm{scan}} \\
&+u^2\left(C_{\mathrm{noise,\,DUT,\,cal}}\right)+u^2\left(C_{\mathrm{noise,\,TS,\,cal}}\right)+u^2\left(C_{3,\,\mathrm{DUT,\,cal}}-C_{3,\,\mathrm{TS,\,cal}}\right)
\end{aligned}
\tag{5.14}
$$

The measurement of the DUT field quantities include measurements only at one distance for each measurement configuration. In order to calculate the uncertainty of the average sound pressure and sound intensity level, the dispersion of the partial surface levels was used [JCG08]:

$$
u^2\left(\overline{L'_{p,\,\mathrm{DUT,\,cal}}}\right)_{\mathrm{scan}}=\frac{\sigma^2\left(L'_{p,\,\mathrm{DUT,\,cal},\,i}\right)_{\mathrm{scan}}}{n_{S\,\mathrm{partial}}}
\tag{5.15}
$$

where $n_{S\,\mathrm{partial}}$ is the number of partial surfaces.

The uncertainty of the DUT sound power level based on sound intensity measurements for the box surface method is:

$$
\begin{aligned}
u^2\left(L_{W,\mathrm{DUT,cal},I}\right)_{\mathrm{discrete}} &= u^2\left(L_{W,\mathrm{TS,cal},I}\right) \\
&+u^2\left[\overline{L_{I,\mathrm{DUT,cal}}}-\overline{L_{I,\mathrm{TS,cal}}}+10\lg\left(\frac{S_{\mathrm{DUT,cal}}}{S_{\mathrm{TS,cal}}}\right)\mathrm{dB}\right]_{\mathrm{scan}} \\
&+u^2\left(\overline{L_{I,\mathrm{DUT,cal}}}\right)_{\mathrm{discrete}}+u^2\left(\overline{L_{I,\mathrm{TS,cal}}}\right)_{\mathrm{discrete}}+u^2\left(C_{3,\mathrm{DUT,cal}}-C_{3,\mathrm{TS,cal}}\right)
\end{aligned}
\tag{5.16}
$$

The only pair of surface and time averaged sound intensity levels of the DUT and the transfer standard is not sufficient to calculate the level difference uncertainty. The uncertainty of the sound intensity level differences for the scanning method was used instead, since the different radii for which it has been calculated can be used as indicative for the box surface method, because it also refers to average levels. For the uncertainty of the sound intensity levels Eq.(5.15) was used.

Accordingly, the formula for the sound pressure levels is:

$$\begin{aligned} u^2\left(L_{W,\,\mathrm{DUT,\,cal,}\,p}\right)_{\mathrm{discrete}} &= u^2\left(L_{W,\,\mathrm{TS,\,cal,}\,p}\right) \\ &+u^2\left[\overline{L_{p,\,\mathrm{DUT,\,cal}}}-\overline{L_{p,\,\mathrm{TS,\,cal}}}+10\lg\left(\frac{S_{\mathrm{DUT,\,cal}}}{S_{\mathrm{TS,\,cal}}}\right)\mathrm{dB}\right]_{\mathrm{scan}} \\ &+u^2\left(\overline{L'_{p,\,\mathrm{DUT,\,cal}}}\right)_{\mathrm{discrete}}+u^2\left(\overline{L'_{p,\,\mathrm{TS,\,cal}}}\right)_{\mathrm{discrete}} \\ &+u^2\left(C_{\mathrm{noise,\,DUT,\,cal}}\right)+u^2\left(C_{\mathrm{noise,\,TS,\,cal}}\right)+u^2\left(C_{3,\,\mathrm{DUT,\,cal}}-C_{3,\,\mathrm{TS,\,cal}}\right) \end{aligned} \tag{5.17}$$

The uncertainties of the sound power level determination of the AEG source as described by Eqs.(5.13), (5.14), (5.16) and (5.17) are shown in Figure 5.13-Figure 5.16. As it may be observed, the most influential partial uncertainty for all cases is the uncertainty of the transfer standard sound power level. The sound pressure level and sound intensity level difference uncertainty is also influential along with the uncertainty of the sound pressure level and sound intensity level. The influence of the background noise and air absorption correction are not influential. The same observations may be made for the air compressor and the vacuum cleaner. All three DUT sound power levels have similar combined uncertainties as it may be seen in Figure 5.17 for the case of sound intensity discrete measurements. The maximum value pro frequency was chosen to derive a representative uncertainty to cover all measured DUT.

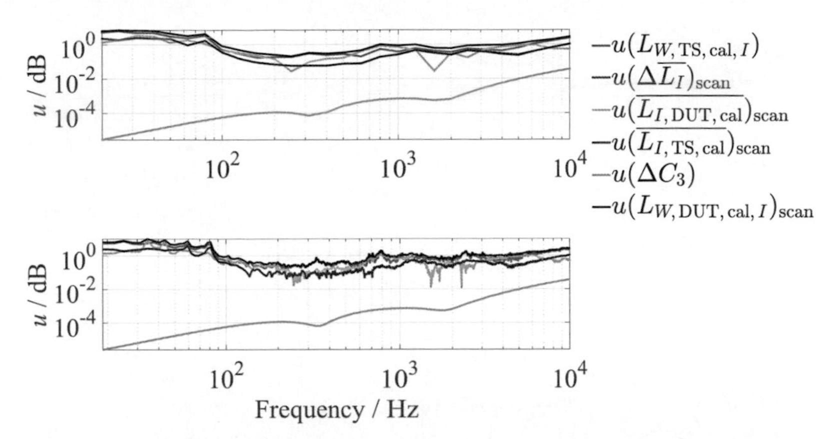

Figure 5.13: Total and partial uncertainties of the AEG source sound power level for the investigation of the measurement surface influence. Sound intensity scanning measurements. Top: one-third octave bands. Bottom: FFT bands.

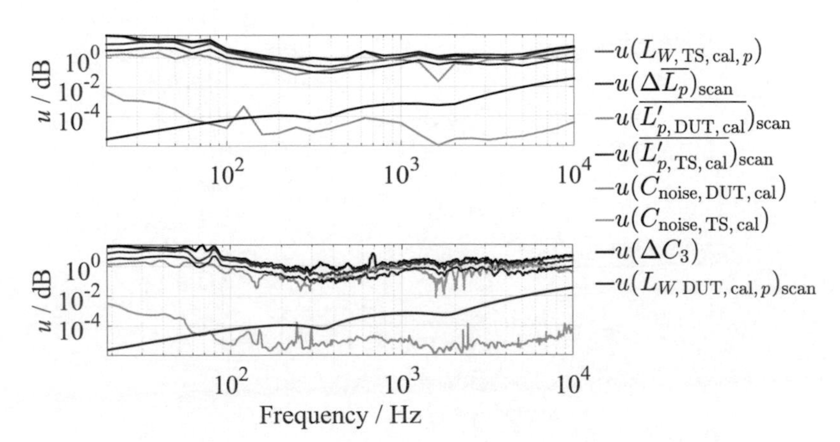

Figure 5.14: Total and partial uncertainties of the AEG source sound power level for the investigation of the measurement surface influence. Sound pressure scanning measurements. Top: one-third octave bands. Bottom: FFT bands.

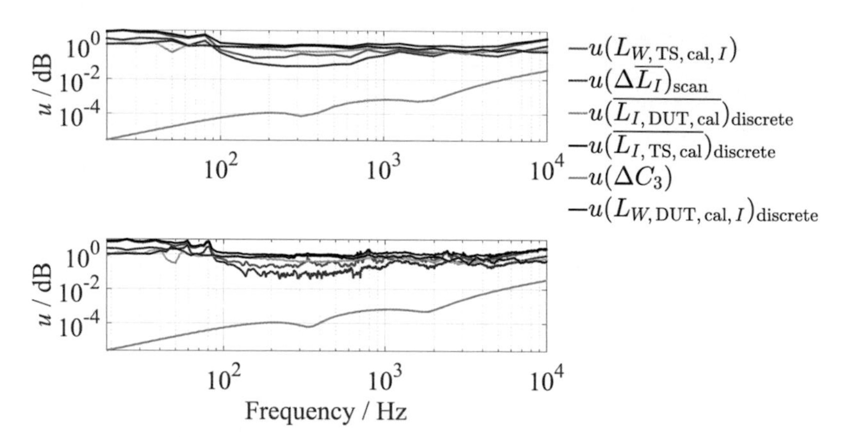

Figure 5.15: Total and partial uncertainties of the AEG source sound power level for the investigation of the measurement surface influence. Sound intensity discrete point measurements. Top: one-third octave bands. Bottom: FFT bands.

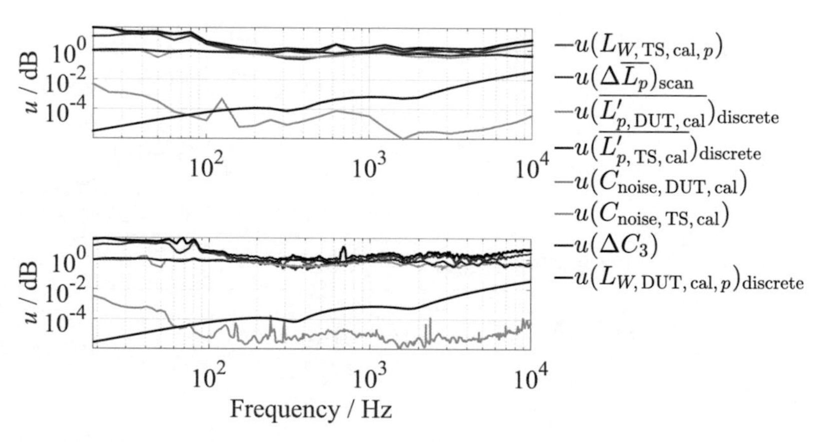

Figure 5.16: Total and partial uncertainties of the AEG source sound power level for the investigation of the measurement surface influence. Sound pressure discrete point measurements. Top: one-third octave bands. Bottom: FFT bands.

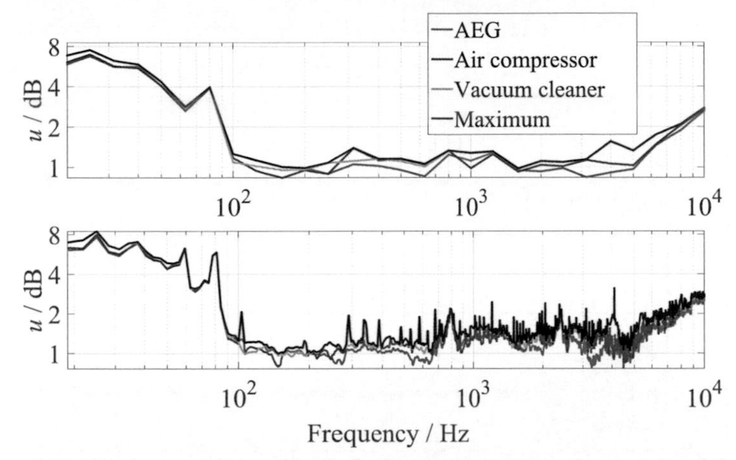

Figure 5.17: Total uncertainty of each device under test sound power level for the investigation of the measurement surface influence and maximum value of all. Sound intensity discrete point measurements. Top: one-third octave bands. Bottom: FFT bands.

For the validation of the calculated uncertainties, the following procedure was followed. The mean value of the directly determined sound power levels and the sound power levels after the substitution method were calculated and the difference between each level and the mean value was produced. This was performed for each source and measurement method. The sound power level differences were compared to literature and the calculated uncertainty, in order to investigate if the differences lie within the expanded uncertainty limits. According to literature, for the sound pressure measurements, the uncertainties of ISO 3745 were chosen [ISO3745]. For the sound intensity measurements, the values of ISO 9614-1 [ISO96141] were chosen for the discrete measurements and ISO 9614-2 [ISO96142] for the scanning measurements. The expanded uncertainty with 95 % confidence interval was calculated according to GUM [JCG08] as $\pm 2u$. The sound power level differences along with the uncertainty limits are presented in Figure 5.18-Figure 5.21. The spiky envelope at high frequencies is due to the air compressor uncertainty (Figure 5.17).

The expanded uncertainty after the substitution method covers the sound power level differences in the whole frequency range in case of sound pressure measurements. At low frequencies, the uncertainty is large due to near field and room modes as previously explained. The existing in literature uncertainty covers the sound power level differences above 90 Hz. The differences corresponding to the directly determined sound power levels lie outside the limits of the existing uncertainty at specific frequencies starting from 700 Hz. The uncertainty based on the substitution method is in general larger than the literature uncertainty.

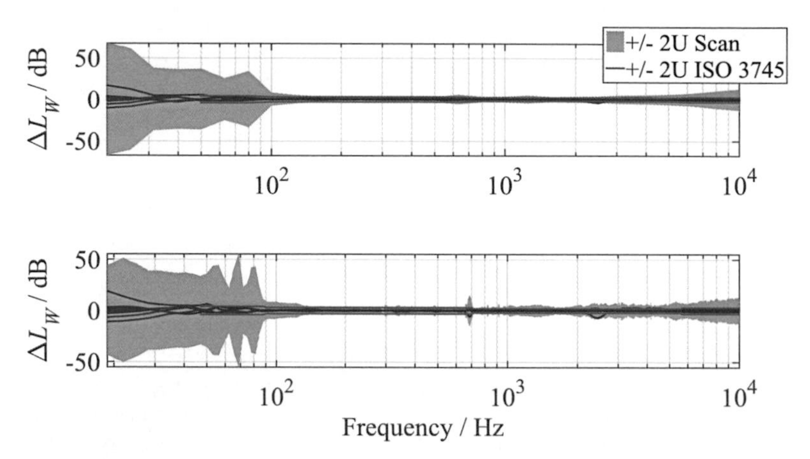

Figure 5.18: Sound power level differences and expanded uncertainty limits for sound pressure scanning measurements. Top: one-third octave bands. Bottom: FFT bands.

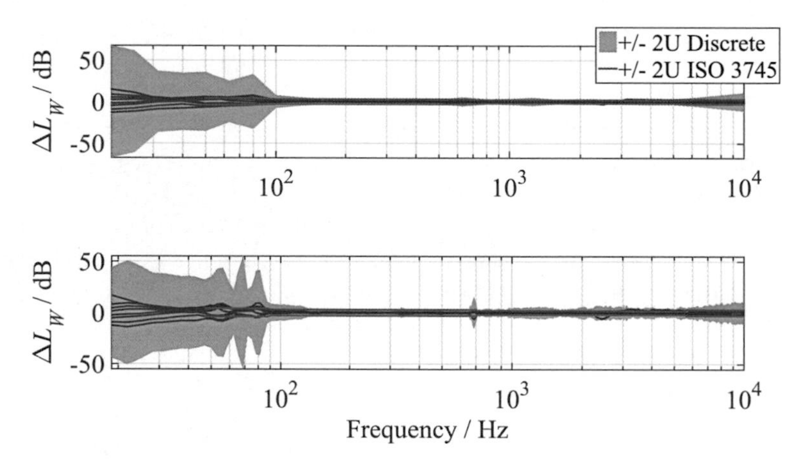

Figure 5.19: Sound power level differences and expanded uncertainty limits for sound pressure discrete point measurements. Top: one-third octave bands. Bottom: FFT bands.

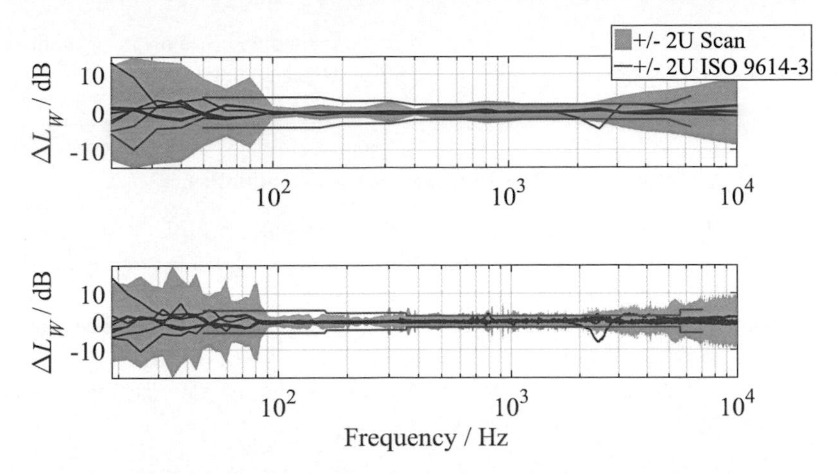

Figure 5.20: Sound power level differences and expanded uncertainty limits for sound intensity scanning measurements. Top: one-third octave bands. Bottom: FFT bands.

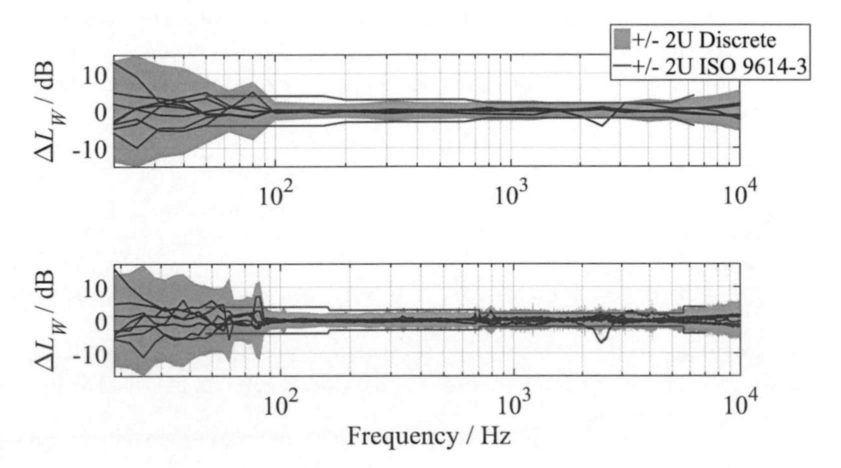

Figure 5.21: Sound power level differences and expanded uncertainty limits for sound intensity discrete point measurements. Top: one-third octave bands. Bottom: FFT bands.

For sound intensity measurements, the proposed uncertainty also covers the sound power level differences except some narrow frequency bands, which may be explained by the 95% confidence interval. Up to 700 Hz the proposed uncertainty is lower than the existing in literature. In overall, it can be stated that the influence of the measurement surface on the sound power determination based on the substitution method can be sufficiently covered by the proposed uncertainty.

5.4.2 Uncertainty of the surrounding environment influence

The uncertainty of the sound power level for the influence of the surrounding environment was also determined. The sound power level of the transfer standard under calibration conditions is given by Eqs.(5.11) and (5.12). The DUT measurements in situ require the correction for the transfer source described by Eq.(4.6) and the calculation of the related uncertainty by Eq.(4.23). Then, the uncertainty of the DUT sound power level after sound intensity measurements is described by:

$$u^2\left(L_{W,\,\text{DUT, in situ},\,I}\right)=u^2\left(L_{W,\,\text{TS, in situ},\,I}\right)+u^2\left(\overline{L_{I,\,\text{DUT, in situ}}}-\overline{L_{I,\,\text{TS, in situ}}}\right)$$
$$+u^2\left(\overline{L_{I,\,\text{DUT, in situ}}}\right)+u^2\left(\overline{L_{I,\,\text{TS, in situ}}}\right) \tag{5.18}$$

The uncertainty of the sound intensity level difference was calculated based on the five differences of all measurement environments. The uncertainty of the sound intensity levels was calculated by the dispersion of the partial sound intensity levels according to Eq.(5.15).

The uncertainty for the case of sound pressure measurements is:

$$u^2\left(L_{W,\,\text{DUT, in situ},\,p}\right)=u^2\left(L_{W,\,\text{TS, in situ},\,p}\right)+u^2\left(\overline{L_{p,\,\text{DUT, in situ}}}-\overline{L_{p,\,\text{TS, in situ}}}\right)$$
$$+u^2\left(\overline{L'_{p,\,\text{DUT, in situ}}}\right)+u^2\left(\overline{L'_{p,\,\text{TS, in situ}}}\right)+u^2\left(C_{\text{noise},\,\text{DUT, in situ}}\right)+u^2\left(C_{\text{noise},\,\text{TS, in situ}}\right) \tag{5.19}$$

The uncertainty of the air absorption correction was omitted since the DUT and transfer source measurements were performed on the same environmental conditions.

As in the case of the measurement surface influence, the total uncertainty is mainly affected by the uncertainty of the in situ sound power level measurement of the transfer source, the uncertainty of the sound pressure level and sound intensity level difference and the uncertainty of the sound pressure and sound intensity. The surrounding environment affects the related uncertainty, which is also different for each DUT. This is shown in Figure 5.24 for the case of sound pressure measurements. To derive a total uncertainty applicable to all surrounding environments, the maximum value per frequency was chosen.

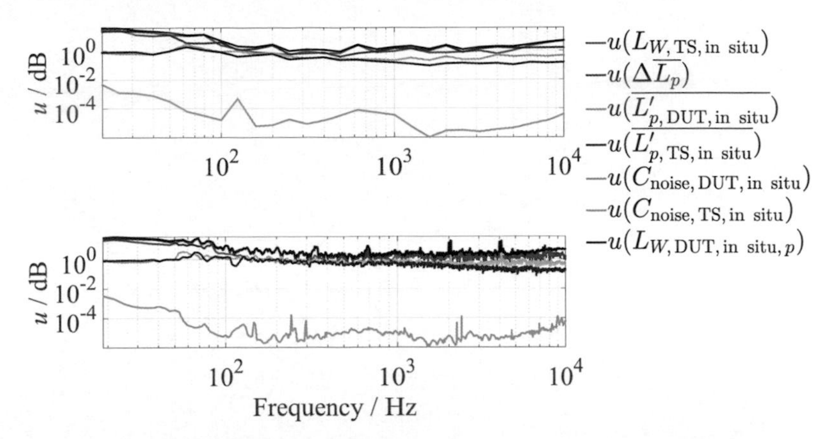

Figure 5.22: Total and partial uncertainties of the device under test sound power level for the investigation of the surrounding environment influence. Sound pressure discrete point measurements of the vacuum cleaner at the hard-walled test room 1. Top: one-third octave bands. Bottom: FFT bands.

The uncertainties of Eq.(5.18) and (5.19) are shown in Figure 5.22 and Figure 5.23 for the hard-walled test room 1.

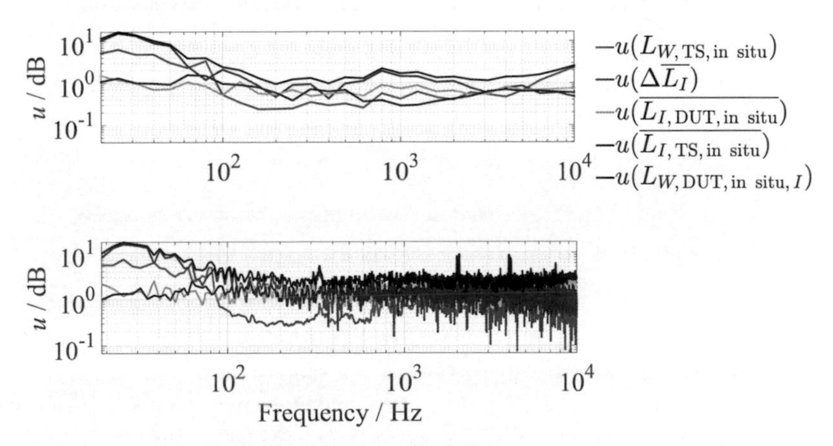

Figure 5.23: Total and partial uncertainties of the device under test sound power level for the investigation of the surrounding environment influence. Sound intensity discrete point measurements of the vacuum cleaner at the hard-walled test room 1. Top: one-third octave bands. Bottom: FFT bands.

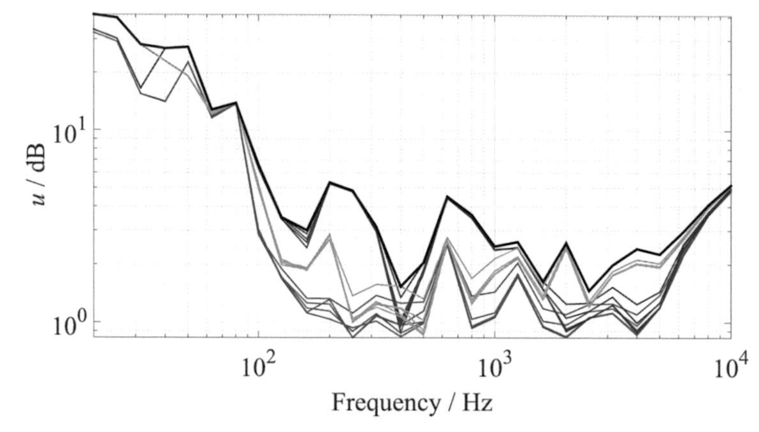

Figure 5.24: Total uncertainty of the AEG source (blue), air compressor (orange red) and vacuum cleaner (yellow) for the investigation of the surrounding environment influence and maximum value of all (black). Sound pressure discrete point measurements.

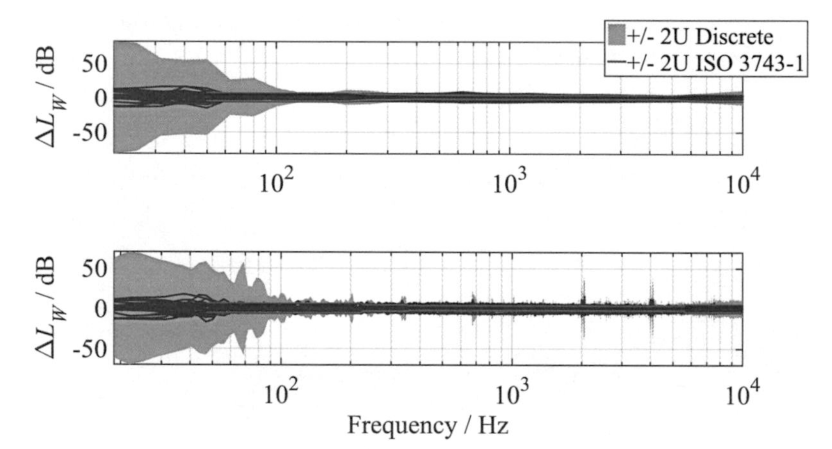

Figure 5.25: Sound power level differences for all surrounding environments and expanded uncertainty limits for sound pressure measurements using the discrete point method. Top: one-third octave bands. Bottom: FFT bands.

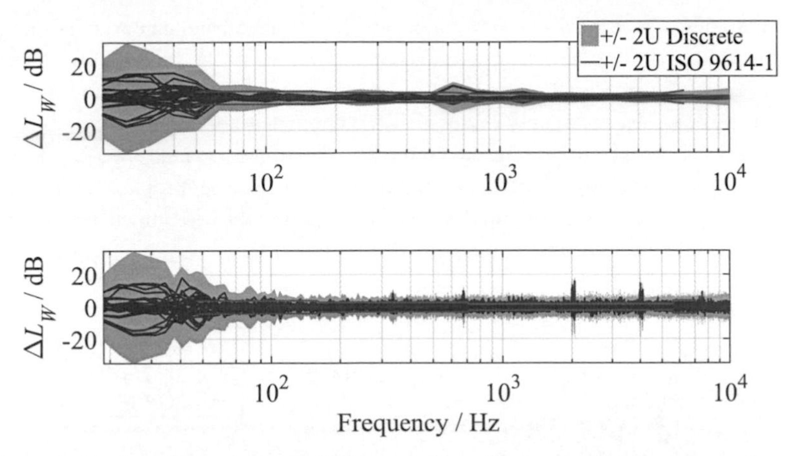

Figure 5.26: Sound power level differences for all surrounding environments and expanded uncertainty limits for sound intensity measurements using the discrete point method. Top: one-third octave bands. Bottom: FFT bands.

The difference between the sound power levels after direct determination and after the substitution method and their mean value was again calculated and compared to the expanded proposed and found in literature uncertainty. For the latter, the values of ISO 3743-1 [ISO37431] were used for the sound pressure measurements and the values of ISO 9614-1 [ISO96141] for the sound intensity measurements. The sound power level differences along with the expanded uncertainties are shown in Figure 5.25 and Figure 5.26. The uncertainty limits for the sound intensity measurements cover the calculated sound power level differences, while this is not the case for the sound pressure measurements. This is attributed to the absence of K_2 correction for the directly calculated sound power levels. Figure 5.27 shows the sound power level differences related only to the substitution method results.

The sound power level differences based on sound pressure measurements and after applying the substitution method are covered by both expanded uncertainties. For frequencies above 70 Hz the ISO 3743-1 [ISO37431] uncertainty has lower values than the proposed one. The latter covers the lower frequencies sufficiently. For the sound intensity measurements, the proposed uncertainty covers all differences for the whole frequency range of interest in both frequency analyses.

In general, it may be concluded that sound power level differences including both direct and after the substitution method results for various measurement surfaces and surrounding environments can be covered by combining the uncertainties currently stated in the ISO series ([ISO37431], [ISO3745], [ISO96141], [ISO96143]) and the uncertainties proposed by this study. For one-third octave band analysis and in the common frequency range, the lowest uncertainty value may be accordingly chosen, to cover the wanted sound power level differences. For frequencies outside the ISO range, the proposed uncertainty values are available. The same is also the case for the FFT analysis. As the analysis used both directly determined sound power levels and levels after the substitution method, the uncertainties are also applicable to each sound power level determination method except for the directly determined levels after sound pressure levels at different surrounding environments.

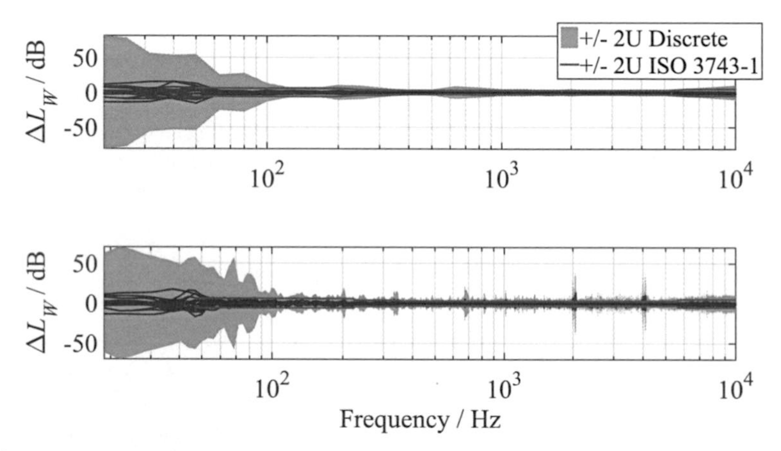

Figure 5.27: Sound power level differences for the substitution method results for all surrounding environments and expanded uncertainty limits for sound pressure discrete point measurements. Top: one-third octave bands. Bottom: FFT bands.

5.5 Dissemination combined uncertainty

The dissemination process includes several applications of the substitution method as it has already been explained. Each time the substitution method is applied a related uncertainty may be derived, allowing a combined uncertainty for the dissemination to be determined. The calculations performed in the previous chapters allow the uncertainty derivation for two major cases: for different measurement surfaces and various surrounding environments. The uncertainty includes also the performance of sound pressure and sound intensity measurements.

The flowchart in Figure 5.28 shows the steps at which the substitution method was applied during the dissemination. Further, the propagation of uncertainty may also be derived in the sense of a relative work [Wit052]. The sound power level of the primary source was determined by laser scanning vibrometer measurements. The sound power level of the transfer source under calibration conditions was determined by the substitution method and sound pressure scanning measurements.

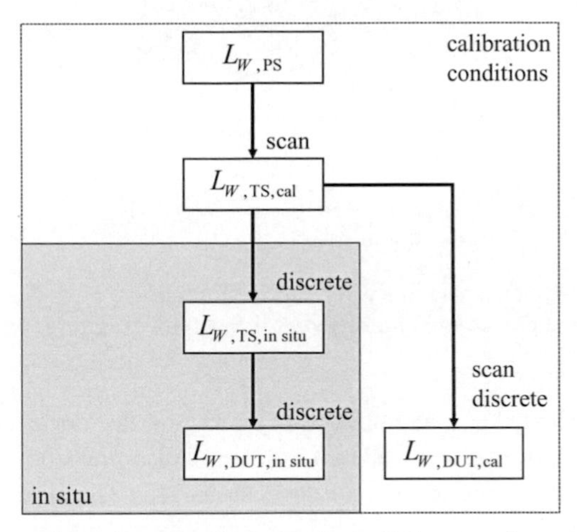

Figure 5.28: Flow chart for the dissemination of the sound power unit and the determination of the related uncertainties.

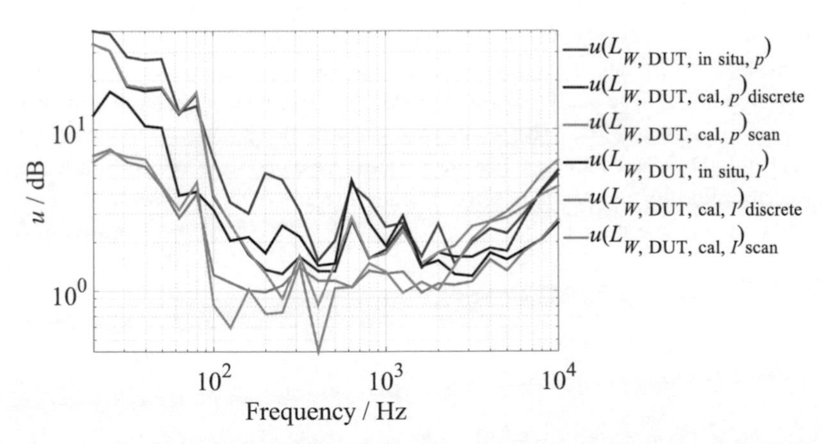

Figure 5.29: Combined uncertainty for the determination of the device under test sound power level for different measurement settings in one-third octave bands.

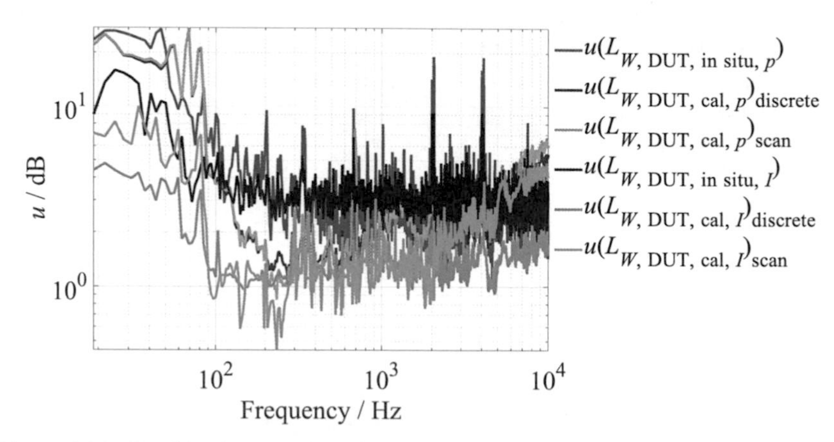

Figure 5.30: Combined uncertainty for the determination of the device under test sound power level for different measurement settings in FFT bands.

In calibration conditions, the sound power level of the device under test was determined by the substitution method for different measurement surfaces and both sound pressure and sound intensity measurements. The sound power level of the transfer source under calibration conditions was corrected and sound pressure and sound intensity measurements at discrete points, led to the sound power level determination of the device under test at various surrounding environments.

The determined uncertainty for all previously mentioned sound power level determination settings is presented in Figure 5.29 and Figure 5.30. By observing the combined uncertainties, it may be said that the sound intensity measurements provided lower uncertainty values compared to the sound pressure measurements, especially below 100 Hz. The in situ measurements have larger uncertainty values than the calibration conditions. The uncertainties have been explicitly determined and can constitute the transparent uncertainty budget required for the dissemination of the unit watt in airborne sound.

6 Conclusions and outlook

The main aim of this study is the establishment of traceability in airborne sound power, which includes the determination of sound power and the related uncertainty. For the determination, a number of influential parameters was investigated focusing on the substitution method, the frequency analysis, the measurements and corrections related to the use of aerodynamic reference sound sources. In the beginning, theoretical calculations were performed to explore the limitations of the substitution method, which is considered to be the fundamental tool for the dissemination of the unit watt from calibration conditions to in situ conditions.

Before the implementation of the substitution method, the model parameters were examined and it was found that the calculation of the sound intensity based on the derivation of the sound pressure reduces the calculation time. It also has better high frequency response, because it is not influenced by the spacer distance, which is the case of the sound intensity calculation based on the sound pressure gradient approximation.

Initially, the substitution method was implemented in free field without reflecting planes. The findings validate the use of the sound intensity for the substitution method, because it enables the determination of the free field sound power level of the unknown source independently of the order of the sources used in the calculations. This is not the case for the substitution implementation using sound pressure where the free field sound power level is determined with small deviations only in the case where both sources are monopoles. The substitution method overestimates the sound power level of the unknown source when the sources are both dipoles or of different order. The overestimation depends on kr and increases by decreasing kr.

The next step of the theoretical calculations included the presence of a reflecting plane, either below or next to the hemispherical measurement surface. The presence of the reflecting plane indicates the intereference of the direct and the reflected sound. For the latter, the propagation of spherical waves was considered along with this of plane waves, which is a simplification of the spherical wave approach. The plane wave approach is straightforward and can be easily implemented. The spherical wave approach on the contrary, is time consuming and requires special attention to computational parameters such as: the number of the measurement surface points, the positioning of the points on the surface and the angle resolution for the reflection on the impedance boundary.

The substitution method is strongly affected by two parameters: the order of the sources and their positioning. If both sources are of the same order, the direct determination of the sound power reveals the overestimation due to the radiation impedance change. This is not the case for the substitution method, which provides sound power levels close to the free field ones. The deviations from the free field level becomes smaller when the sound intensity is used for the implementation of the substitution method. The deviation from the free field sound power level depends on the positioning of the known and the unknown source. For small deviations, the horizontal translation of the unknown source can be larger than the related vertical translation. Both direct and substitution method provide overestimated sound power levels in case the sources are of different order.

In overall, the sound intensity could preferably be used for the substitution method and the sources should ideally be of the same order. The substitution of a source over a reflecting plane could be modelled by FEM as a next step to the theoretical investigation of the substitution method. Another study would be the inclusion of two reflecting planes in a model.

In the first experimental part of the study, aerodynamic reference sound sources were investigated for transfer standard qualification. For the substitution method implementation, a scanning apparatus was specially designed. It enabled the surface averaged sound pressure and sound intensity to be measured at different radii. The application of the substitution method has among others, the advantage to provide cancellation to effects, which would otherwise affect the measurements and the related uncertainty. An important cancellation is this of the reflections from the scanning apparatus metallic body provided that both sources are measured on the same radius. It was found that the repeatability of the scanning apparatus measurements is higher than this of the spiral method, which is used in current reference sound source qualification procedures [ISO6926]. The frequency analysis in both broad and narrow bands, provided qualitative results to be used for the broadening of the application of the reference sound sources to determine the sound power of sources with tonal characteristics. The development of reference sound sources of such characteristics could be a field for further research. An additional important feature of new sources would be the constant volume flow, which would make the substitution method foundations more robust. Since the rotating fan of the aerodynamic reference sound sources is related to wind generation, another study would be about the removal of the wind influences based on cross-correlation measurements.

The second experimental investigation focused on the sound generation by aerodynamic reference sound sources, which is proportional to environmental pressure, ambient temperature and fan rotation speed. The dissemination of the unit watt relates the sound power under calibration conditions to in situ conditions. For the conditions transition, a correction is required to be applied to the sound power level of the transfer standards under calibration conditions. Special attention was paid to the correction factors estimation. The small range of the environmental pressure values due to the unaltered altitude could not lead to robust results concerning the experimental correction for changes due to atmospheric pressure variations. Theoretical values were used instead, since no great deviations from these are expected to be seen by measurements. Variations in the measurement site altitude could improve the results. Sound pressure measurements in various ambient temperatures led to a low frequency limited correction factor due to the influence of background noise during outdoor measurements. Extrapolation of the available values was the basis for the missing low frequency part assuming not significant chance of the source radiation order. Indoor measurements while varying the ambient temperature would provide entirely experimental results. Variations in the fan rotation speed included two different alternating current frequencies. The results revealed that a unique correction may be applied to both frequencies.

The most prominent result of the determination of the above factors is that the radiation order of the reference sound sources depends on frequency. Most apparent is the increasingly high order behaviour at high frequencies while varying the fan rotation speed. The same findings were further evaluated by spherical harmonics decomposition. The full sphere equations used, because the mirroring on the hemianechoic room floor was assumed. It would be of interest to apply the spherical harmonics transform by changing the equations solid angle limits to correspond to a hemisphere. Apparently, the reference sound sources cannot be uniquely described in terms of sound radiation order, but this must be done related to frequency. The verification of the proposed correction was performed by comparison to an existing one, with the former yielding better results in terms of deviation from the mean value of the corrected sound power level.

The high influence of the near field at low frequencies was also explored by theoretical and experimental calculations. It was found that the measurement results describe better the occurring physical phenomena, because they also include the influence of the room remaining modes and the room volume, which are also apparent at low frequencies.

The substitution method was applied for the sound power determination of equipment and machinery in situ using both sound pressure and sound intensity measurements. The influence of different measurement surfaces was explored in terms of scanning and discrete method. Measurements also included the sound power determination at various surrounding environments. It was shown that the directly determined sound power is larger up to nearly 25 dB compared to the free field sound power at low frequencies. The sound intensity results are less prone to low frequency effects than the sound pressure results. For the latter, the K_2 correction should be included in the case of different surrounding environments. The spread of the sound power levels after the substitution method is smaller than this of the directly determined sound power levels.

Apart from the sound power determination, another research field was the determination of the uncertainty at each step of the substitution method application. This way, a combined uncertainty can be finally derived. During the determination of the sound power of the transfer source under calibration conditions, the most influential uncertainty parameters are the low frequency effects (near field and room remaining influences). Strongly contributing are also the statistical spreads of the sound pressure averaged levels of both the primary and the transfer source. The former is also influenced by background noise. An improved version of the primary source is expected to reduce the related uncertainties. The use of windscreens also adds an uncertainty component at high frequencies.

In the next step of the dissemination process, which is the correction of the transfer source sound power level from calibration conditions to in situ conditions, the related uncertainty is mostly influenced by the uncertainty of the sound power determination in calibration conditions. On the other hand, the uncertainty contribution of the correction factors is relatively small.

The uncertainty of the sound power level of real sources depends on the measurement environment. In calibration conditions, the most influential component is the uncertainty of the sound power level of the transfer source. Additionally, the sound pressure or sound intensity level difference uncertainties also contribute to the overall uncertainty to a significant extend. When the uncertainty is related to different surrounding environments, the component related to the sound power level of the transfer standard becomes not influential. The overall uncertainty is determined by the uncertainties of the sound pressure or sound intensity level differences.

By comparing the use of sound pressure and sound intensity, the sound power determination based on sound intensity under calibration conditions utilizing the scanning method, yields smaller uncertainty than sound pressure. The lower uncertainty is also seen when the comparison is against measurements at discrete points especially at low frequencies. Sound intensity measurements have smaller uncertainty when the sound power determination focuses on various surrounding environments. The uncertainty calculation focused also on the determination of a budget for all used real sources. The final result strongly depends on the spectral characteristics of the sources especially when tonal components are apparent.

An intriguing point for further examination would be the assembly of transfer sources with varying directivity and order. This way the matching between transfer and real source could lead to better substitution results. This kind of substitution would require the radiation pattern of both the transfer and the real source, which could be provided by the spherical harmonics transform.

To summarize the contribution of the thesis to the sound power determination, Table 6.1 contains the partial studies performed towards the dissemination of the sound power unit and the estimation of the related uncertainty. Initially, a theoretical investigation of the substitution method was performed, revealing strong dependence to the radiation order of the sources used for the substitution implementation and their relative positioning. Secondly, the sound power determination of transfer sources under calibration conditions was implemented, where a scanning apparatus enabled the surface averaged sound pressure or sound intensity level by the physical scan of a hemisphere. The third study focused on the required correction for the sound power determination of transfer sources in situ due to changes in atmospheric pressure, ambient temperature and fan rotation speed. Lastly, the application of the proposed method was implemented to real sources with focus on the influence of the measurement surface and the surrounding environment.

By concluding, the present study may be used as the tool for a detailed and easily implementable dissemination procedure for the sound power determination and its related uncertainty in airborne sound. The investigation of transfer standards of different types provided conclusive results for this kind of sources. Based on the above, the sound power determination of machinery and equipment along with the related uncertainty has been achieved, again by a procedure, which covers all measured sources. It must be mentioned once more, that the real sources had different spectral contents. The comparison of the determined uncertainty to the uncertainty given in the current sound power standards, revealed that the proposed method may provide an extension of the uncertainty frequency range to frequencies below the current limits. The combination of both uncertainties may be used to improve the uncertainty budget for the sound power determination using the substitution method. The results of the study may contribute to a new standardized sound power determination procedure, which would provide the free field sound power using the substitution method. An evolution of the substitution method would be the inclusion of sound intensity measurements apart from sound pressure measurements. The proposed analysis provides also a transparent uncertainty budget, which explicitly decomposes the uncertainty contributors at each stage of the sound power determination, which is missing from the up-to-date related literature.

Table 6.1: Summarizing table for the contribution of the thesis to sound power determination.

<table>
<tr><td colspan="5">Focus: substitution method</td></tr>
<tr><td colspan="5">Study: theoretical investigation</td></tr>
<tr><td rowspan="10"></td><td rowspan="2">Field quantity</td><td colspan="3">Sound pressure</td></tr>
<tr><td colspan="3">Sound intensity</td></tr>
<tr><td>Frequency resolution</td><td colspan="3">Broad and narrow band</td></tr>
<tr><td>Known source</td><td>Monopole</td><td>Dipole</td><td>Monopole</td></tr>
<tr><td>Unknown source</td><td>Monopole</td><td>Dipole</td><td>Dipole</td></tr>
<tr><td rowspan="3">Sound field</td><td colspan="3">Free</td></tr>
<tr><td>Reflecting floor</td><td colspan="2" rowspan="2">For reflection: plane and spherical wave approach</td></tr>
<tr><td>Absorbing side wall</td></tr>
<tr><td colspan="4">Vertical/horizontal translation of the unknown source</td></tr>
<tr><td colspan="4"></td></tr>
<tr><td colspan="5">Focus: calibration of transfer sources</td></tr>
<tr><td colspan="5">Study: sound power determination of transfer sources under calibration conditions</td></tr>
<tr><td rowspan="4"></td><td rowspan="2">Field quantity</td><td colspan="3">Sound pressure</td></tr>
<tr><td colspan="3">Sound intensity</td></tr>
<tr><td>Frequency resolution</td><td colspan="3">Broad and narrow band</td></tr>
<tr><td colspan="4">Scan over physical hemisphere</td></tr>
<tr><td colspan="5">Focus: transition from calibration to in situ conditions</td></tr>
<tr><td colspan="5">Study: sound power determination of transfer sources in situ</td></tr>
<tr><td rowspan="5"></td><td rowspan="2">Field quantity</td><td colspan="3">Sound pressure</td></tr>
<tr><td colspan="3">Sound intensity</td></tr>
<tr><td>Frequency resolution</td><td colspan="3">Broad and narrow band</td></tr>
<tr><td colspan="4">Correction due to changes in atmospheric pressure, ambient temperature and fan rotation speed</td></tr>
<tr><td colspan="4">Transfer source radiation order investigation based on near field effects and spherical harmonics transform</td></tr>
<tr><td colspan="5">Focus: application to real sources</td></tr>
<tr><td colspan="5">Study: sound power determination of real sources in situ</td></tr>
<tr><td rowspan="4"></td><td rowspan="2">Field quantity</td><td colspan="3">Sound pressure</td></tr>
<tr><td colspan="3">Sound intensity</td></tr>
<tr><td>Frequency resolution</td><td colspan="3">Broad and narrow band</td></tr>
<tr><td colspan="4">Investigation for the influence of measurement surface and surrounding environment</td></tr>
<tr><td colspan="5">Final result: dissemination implementation + transparent uncertainty budget</td></tr>
</table>

7 Literature

[Age93] Agerkvist F. T. and Jacobsen F. *Sound power determination in reverberation rooms at low frequencies*. J. Sound Vib. 1993, Vol. 166(1), pp. 179-190.

[All79] Allen J. B. and Berkley D. A. *Image method for efficiently simulating room-acoustics*. J. Acoust. Soc. Am. 1979, Vol. 65(4), pp. 943-950.

[Ant08] An-tze S. *Improvement of sound power level nearfield measurement on electrical machine in situ*. Internoise Proceedings. 1980, pp. 1071-1078.

[Bal82] Ballagh K. O. *Limitations to the measurement of sound power by the reference sound source method*. J. Acoust. Soc. Am. 1982, Vol. 72(5), pp. 1637-1639.

[Bel01] Bell S. *A beginner's guide to uncertainty of measurement*. National Physical Laboratoy Measurement Good Practice Guide. No11, 2001.

[Ber17] Berzborn, M., et al. *The ITA-Toolbox: An open source MATLAB Toolbox for acoustic measurements and signal processing*. DAGA Proceedings. 2017, pp. 222-225.

[Bie03] Bies D. A. and Hansen C. H. *Engineering noise control*. London: Spon Press, 2003.

[Bie93] Bies D. A. and Bridges G. *Sound power determination in the geometric near field of a source by pressure measurements alone*. Annual Conference of the Australian Acoustical Society Proceedings. 1993, pp. 54-58.

[Bir03] Birch K. *Estimating uncertainties in testing*. National Physical Laboratory Measurement Good Practice Guide. No36, 2003.

[Brü13] Brüel&Kjær. *Reference sound source Type 4204, User Manual, English.* 2013.

[Brü76] Brüel&Kjær. *Technical Review.* No 4, 1976.

[Brü822] Brüel&Kjær. *Condenser microphones and microphone preamplifiers for acoustic measurements, Data handbook.* 1982.

[Brü87] Brüel&Kjær. *Technical Review.* No 3, 1987.

[Cal07] Caligiuri L. M. *The evaluation of uncertainty in environmental acoustic measurements according to the ISO 'Guide'*. Noise Control Eng. J. 2007, Vol. 55(1), pp. 116-132.

[Cam08] Campanella A. *A review of reference sound sources for sound power measurements*. Sound Vib. 2008, 2, pp. 7-9.

[Cam91] Campanella A. J. *Improved reference sound source for field calibration of sound intensity and sound power measurements*. Internoise Proceedings. 1991, pp. 961-964.

[Cam96] Campanella A. J. *Calibration of reference sound sources - A US historical perspective*. Noise-Con Proceedings. 1996, pp. 931-936.

[Car07] Carletti E. *Towards a harmonized procedure for the declaration of sound power levels within Directive 2000/14/EC*. Noise Control Eng. J. 2007, Vol. 55(1), pp. 12-19.

[Car16] Carletti E. and Pedrielli F. *Outdoor machinery: a reliable statistical approach for a new noise labelling based on current noise emission marking data*. ICSV23 Proceedings. 2016, pp. 4302-4309.

[Cor94] Corrêa F. R. P. and Araújo, M. A. N. *Interference effects on sound power measurement*. Internoise Proceedings. 1994, pp. 1793-1796.

[Cor96] Corrêa F. P. de R. *Acoustic center determination on anechoic half-space*. Appl. Acoust. 1996, Vol. 48(4), pp. 357-361.

[Cun06] Cunefare K. A., Badertscher J. and Wittstock V. *On the qualification of anechoic chambers; Issues related to signals and bandwidth*. J. Acoust. Soc. Am. 2006, Vol. 120(2), pp. 820-829.

[Dir14] 2000/14/EC Directive. *Noise - Equipment for use outdoors.*

[Dir30] 2010/30/EU Directive. *Energy Labelling Directive.*

[Dir42] 2006/42/EC Directive. *New Machinery Directive.*

[Dra11] Dragonetti R., Ianniello C. and Romano R. A. *Calibration of a compound dodecahedron loudspeaker to implement a reference sound source*. Forum Acusticum Proceedings. 2011, pp. 1869-1874.

[Fah95] Fahy F. *Sound intensity.* London: E & FN Spon, 1995.

[Fah98] Fahy F. and Walker J. *Fundamentals of noise and vibration.* London : Spon Press, 1998.

[Fin93] Finke H.-O. and Bethke C. *Different results of sound power measurements as a consequence of standardization?* Internoise Proceedings. 1993, pp. 335-340.

[Fra77] François P. *Characteristics and calibration of reference sound sources*. Noise Control Eng. 1977, Vol. 9 (1), pp. 6-15.

[Gol10] Goldsmith M. *A beginner's guide to measurement*. National Physical Laboratory Good Practice Guide. No118, 2010.

[Gue16] Guevara B. J. y S., et al. *Guide for manufacturers on how to report noise emission in instruction manuals and other literature in accordance with Machinery Directive 2006/42/EC and Outdoor Noise Directive 200/14/EC*. Federal Institute for Occupational Safety and Health Report. 2016.

[Hag11] Hagai I. B., et al. *Acoustic centering of sources measured by surrounding spherical microphone arrays*. J. Acoust. Soc. Am. 2011, Vol. 130(4), pp. 2003-2015.

[Han92] Hanes P. *Measurement uncertainties in the determination of machinery sound power levels*. Internoise Proceedings. 1992, pp. 285-291.

[Har59] Hardy H. C. *Standard mechanical noise sources*. Noise Control. 1959, Vol. 5, pp. 22-25.

[Hes08] Hessler G. F., et al. *Experimental study to determine wind induced noise and wind screen attenuation effects on microphone response for environmental wind turbine and other applications*. Noise Control Eng. J. 2008, Vol. 56(4), pp. 300-309.

[Hic90] Hickling R. *Narrow-band indoor measurement of the sound power of a complex mechanical noise source*. J. Acoust. Soc. Am. 1990, Vol. 87(3), pp. 1182-1191.

[Hic97] Hickling R., Lee P. and Wei W. *Investigation of integration accuracy of sound-power measurement using an automated sound-intensity system*. Appl. Acoust. 1997, Vol. 50(2), pp. 125-140.

[Hig93] Higginson R. F. and Hanes P. *Measurement uncertainties in determination of noise emission: a review*. Noise Control Eng. 40(2), 1993, pp. 173-178.

[Hol77] Holmer C. I. *Investigation of procedures for estimation of sound power in the free field above a reflecting plane*. J. Acoust. Soc. Am. 1977, Vol. 61(2), pp. 465-475.

[Hü011] Hübner G. and Wittstock V. *Investigations of the sound power of aerodynamic sources as a function of static pressure*. Internoise Proceedings. 2001, pp. 2240-2243.

[Hü731] Hübner, G. *Analysis of errors in measuring machine noise under free-field conditions*. J. Acoust. Soc. Am. 1973, Vol. 54(4), pp. 967-977.

[Hü771] Hübner, G. *Qualification procedures for free-field conditions for sound-power determination of sound sources and methods for the determination of the appropriate environmental correction*. J. Acoust. Soc. Am. 1977, Vol. 61(2), pp. 456-464.

[Hü772] Hübner, G. *Influence of static pressure on noise generation of aerodynamic sources.* ICA Proceedings. 1977, p. 155.

[Hü801] Hübner, G. *Is the sound power defined by ISO/TC43 independent of specific environmental conditions?* ICA Proceedings. 1980, pp. M-6.5.

[Hü981] Hübner G. and Wittstock V. *Investigations of the sound power generation of solid cylinders moving in gases under different static pressures - first results*. Euronoise Proceedings. 1998, pp. 865-870.

[Hü992] Hübner G. *Sound power related to normalized meteorological conditions*. Internoise Proceedings. 1999, pp. 1529-1534.

[IEC14] IEC 61260:2014. *Electroacoustics - Octave-band and fractional-octave-band filters - Part 1: Specifications.*

[Ing51] Ingard U. *On the reflection of a spherical sound wave from an infinite plane*. J. Acoust. Soc. Am. 1951, Vol. 23(3), pp. 329-335.

[ISO3740] ISO/FDIS 3740:2017. *Acoustics - Determination of sound power levels of noise sources - Guidelines for the use of basic standards.*

[ISO3741] ISO 3741:2010. *Acoustics - Determination of sound power levels and sound energy levels of noise sources using sound pressure - Precision methods for reverberation test rooms.*

[ISO37431] ISO 3743-1:2010. *Acoustics - Determination of sound power levels and sound energy levels of noise sources using sound pressure - Engineering methods for small movable sources in reverberant fields - Part 1: Comparison method for a hard-walled test room.*

[ISO37432] ISO 3743-2:2018. *Acoustics - Determination of sound power levels of noise sources using sound pressure - Engineering methods for small, movable sources in reverberant fields - Part 2: Methods for special reverberation test rooms.*

[ISO3744] ISO 3744:2010. *Acoustics - Determination of sound power levels and sound energy levels of noise sources using sound pressure - Engineering methods for an essentially free field over a reflecting plane.*

[ISO3745] ISO 3745:2012. *Acoustics - Determination of sound power levels and sound energy levels of noise sources using sound pressure - Precision methods for anechoic rooms and hemi-anechoic rooms.*

[ISO3746] ISO 3746:2010. *Acoustics - Determination of sound power levels and sound energy levels of noise sources using sound pressure - Survey method using an enveloping measurement surface over a reflecting plane.*

[ISO3747] ISO 3747:2010. *Acoustics - Determination of sound power levels and sound energy levels of noise sources using sound pressure - Engineering/survey methods for use in situ in a reverberant environment.*

[ISO6926] ISO 6926:2016. *Acoustics - Requirements for the performance and calibration of reference sound sources used for the determination of sound power levels.*

[ISO8000] ISO 80000-8:2007. *Quantities and units - Part 8: Acoustics.*

[ISO9613] ISO 9613-1:1993. *Acoustics - Attenuation of sound during propagation outdoors - Part 1: Calculation of the absorption of sound by the atmosphere.*

[ISO96141] ISO 9614-1:1993. *Acoustics - Determination of sound power levels of noise sources using sound intensity - Part 1: Measurement at discrete points.*

[ISO96142] ISO 9614-2:1996. *Acoustics - Determination of sound power levels of noise sources using sound intensity - Part 2: Measurement by scanning.*

[ISO96143] ISO 9614-3:2002. *Acoustics - Determination of sound power levels of noise sources using sound intensity - Part 3: Precision method for measurement by scanning.*

[Jac07] Jacobsen F. *On the uncertainty in measurement of sound power using sound intensity*. Noise Control Eng. J. 2007, Vol. 55(1), pp. 20-28.

[Jac09] Jacobsen F. and Molares A. R. *Sound power emitted by a pure-tone source in a reverberation room*. J. Acoust. Soc. Am. 2009, Vol. 126(2), pp. 676-684.

[Jac94] Jacobsen F. *A note on measurement of sound intensity with windscreened probes*. Appl. Acoust. 1994, Vol. 42(1), pp. 41-53.

[Jac97] Jacobsen F. *An overview of the sources of error in sound power determination using the intensity techique*. Appl. Acoust. 1997, Vol. 50(2), pp. 155-166.

[JCG08] JCGM 100. *Evaluation of measurement data - Guide to the expression of uncertainty in measurement*. 2008.

[JCG12] JCGM 200. *International vocabulary of metrology - Basic and general concepts and associated terms*. 2012.

[Jon86] Jonasson H. G. *Measurements with reference sources in the ISO 3740 series*. Internoise Proceedings. 1986, pp. 1371-1374.

[Jon88] Jonasson H. G. *Accurate sound power measurements using a reference sound source*. Internoise Proceedings. 1988, pp. 1129-1134.

[Juh06] Juhl P. and Jacobsen F. *A numerical investigation of the influence of windscreens on measurement of sound intensity*. J. Acoust. Soc. Am. 2006, Vol. 119(2), pp. 937-942.

[Kir06] Kirbaş C., et al. *Primary sound power sources for the realisation of the unit Watt in airborne sound.* Internoise Proceedings. 2016, pp.6737-6748.

[Kri83] Krishnappa G. *Acoustic intensity in the nearfield of two interfering monopoles*. J. Acoust. Soc. Am. 1983, Vol. 74(4), pp. 1291-1294.

[Kri87] Krishnappa G. *Sound intensity in the nearfield of a point source over a hard reflecting plane*. J. Acoust. Soc. Am. 1987, Vol. 82(2), pp. 667-678.

[Kut79] Kuttruff H. *Room acoustics.* Essex: Applied science publishers Ltd., 1979.

[Lav92] Laville F. and Nicolas J. *A computer simulation of sound power determination using two-microphone sound intensity measurements*. J. Acoust. Soc. Am. 1992, Vol. 91(4), pp. 2042-2055.

[LiK97] Li K. M., Taherzadeh S. and Attenborough K. *Sound propagation from a dipole source near an impedance plane*. J. Acoust. Soc. Am. 1997, Vol. 101(6), pp. 3343-3352.

[LiY96] Li Y. L. and White M. J. *Near-field computation for sound propagation above ground-using complex image theory*. J. Acoust. Soc. Am. 1996, Vol. 99(2), pp. 755-760.

[Loy07] Loyau T. *Determination of sound power levels using sound pressure: The uncertainties related with the measurement surface and the number of microphones*. Noise Control Eng. J. 2007, Vol. 55(1), pp. 89-97.

[Lub74] Lubman D. *Precision of reverberant sound power measurements*. J. Acoust. Soc. Am. 1974, Vol. 56(2), pp. 523-533.

[MAB16] Boucher M. A., Pluymers B. and Desmet W. *Interference effects in phased beam tracing using exact half-spaced solutions*. J. Acoust. Soc. Am. 2016, Vol. 140(6), pp. 4204-4212.

[MAr14] Aretz M., Dietrich P. and Vorländer M. *Application of the mirror source method for low frequency sound prediction in rectangular rooms*. Acta Acust. United Ac. 2014, Vol. 100(2), pp. 306-319.

[Max12] Maximov G. A., Larichev V. A. and Khoroshenkov K. V. *The intensity of the field generated by a point source and reflected from a rough surface*. Acoust. Phys. 2012, Vol. 58(2), pp. 139-146.

[Mec891] Mechel F. P. *Analysis of spherical wave propagation over absorbing ground.* Annual Conference of the Institute of Acoustics Proceedings. 1989, Vol. 11(5), pp. 231-256.

[Mec892] Mechel F. P. *Schallabsorber, Band I, Äußere Schallfelder-Wechselwirkungen.* Stuttgart: S. Hirzel, 1989.

[Nob99] Nobile, M. A. and Shaw, J. A. *The cylindrical microphone array for determining sound power levels of noise sources*. Internoise Proceedings. 1999, pp. 1535-1540.

[Pat93] Paterson A. L. T. *On the mathematics of estimating sound power by the scanning method*. J. Sound Vib. 1993, Vol. 161(2), pp. 241-250.

[Pol15] Pollow M. *Directivity patterns for room acoustical measurements and simulations. RWTH Aachen PhD Thesis*. Berlin: Logos, 2015.

[Pop86] Pope J. *Intensity measurements for sound power determination over a reflecting plane*. Internoise Proceedings. 1986, pp. 1115-1120.

[Pop89] Pope J. *Qualifying intensity measurements for sound power determination*. Internoise Proceedings. 1989, pp. 1041-1046 .

[Pro89] Probst W. *Numerical simulation of the determination of the sound power level for machines. The angle error using the surface envelope method*. Acustica. 1989, Vol. 68(2), pp. 150-156.

[Pro93] Probst W. *Measurement of the noise emission of machines with a reference sound source*. Internoise Proceedings. 1993, pp. 373-376.

[Rus80] Russell G. A. *Error propagation analysis of four sound-power measurement techniques*. J. Acoust. Soc. Am. 1980, Vol. 67(2), pp. 663-665.

[Seh96] Sehrndt G. A. *Remarks on the true value of the sound power level*. Internoise Proceedings. 1996, pp. 2703-2706.

[Sha15] Shabtai N. R. and Vorländer M. *Acoustic centering of sources with high-order radiation patterns*. J. Acoust. Soc. Am. 2015, Vol. 137(4), pp. 1947-1961.

[Shi88] Shirahatti U. S., Crocker M. J. and Raju P. K. *Finite difference approximation errors in sound intensity estimates of interfering sources*. J. Acoust. Soc. Am. 1988, Vol. 84(2), pp. 629-638.

[Shi92] Shirahatti U. S. and Crocker M. J. *Two-microphone finite difference approximation errors in the interference fields of point dipole sources*. J. Acoust. Soc. Am. 1992, Vol. 92(1), pp. 258-267.

[Sim04] Simmons D., Jobling B. and Payne R. *Acoustic parameters and uncertainties associated with determining sound power level in hemi-anechoic rooms*. National Physical Laboratory Report. DQL-AC 007, 2004.

[Som09] Sommerfeld A. *Über die Ausbreitung der Welle in der drahtlosen Telegraphie*. Ann. Phys. 1909, Vol. 333, pp. 665-736.

[Str18] Struck C. J. *Measurement uncertainty and its applications to acoustical standards*. Proc. Mtgs. Acoust. 2017, Vol. 31(1), p. 032001.

[Suh99] Suh J. S. and Nelson P. A. *Measurement of transient response of rooms and comparison with geometrical acoustic models*. J. Acoust. Soc. Am. 1999, Vol. 101(6), pp. 2304-2317.

[Suz07] Suzuki H., Nakamura M. and Tichy J. *An accuracy evaluation of the sound power measurements by the use of the sound intensity and the sound pressure methods*. Acoust. Sci. & Tech. 2007, Vol. 28(5), pp. 319-327.

[Suz93] Suzuki S. *Study on the reference sound source*. Internoise Proceedings. 1993, pp. 427-430.

[Tac89] Tachibana H. and Yano H. *Changes of sound power of reference sound sources influenced by boundary conditions measured by the sound intensity technique*. Internoise Proceedings. 1989, pp. 1009-1014.

[Tac96] Tachibana H., Yano H. and Koyasu M. *Acoustic measurements using an impulsive reference source*. Internoise Proceedings. 1996, pp. 2673-2677.

[Tak00] Takazawa A., et al. *Development of a loud speaker type reference sound source*. Internoise Proceedings. 2000, pp. 1983-1989.

[Tay94] Taylor B. N. and Kuyatt C. E. *Guidelines for evaluating and expressing the uncertainty of NIST measurement results*. National Institute of Standards and Technology Technical Note. 1297, 1994.

[Tho76] Thomasson S.-I. *Reflection of waves from a point source by an impedance boundary*. J. Acoust. Soc. Am. 1976, Vol. 59(4), pp. 780-785.

[Toh87] Tohyama M., Imai A. and Tachibana H. *Standard deviation in the power measurement of sound sources by the free field method*. J. Sound Vib. 1987, Vol. 114(1), pp. 121-127.

[Toh88] Tohyama M., Imai A. and Tachibana H. *Standard deviation in power measurements on small measurement surfaces with use of the free field method*. J. Sound Vib. 1988, Vol. 121(2), pp. 211-220.

[Toh90] Tohyama M. *Biased errors in sound power measurement of a source*. Internoise Proceedings. 1990, pp. 673-676.

[Tsu14] Tsuei K.-Y., et al. *Research on calibration technology for reference sound source and its application*. Internoise Proceedings. 2013, pp. 671-678.

[Und18] Underwood S. H. and Wang L. M. *A round robin study of sound power measurement methods to determine reproducibility and bias*. Internoise Proceedings. 2018, pp. 4674-4681.

[van18] van der Linden P. J. G. and Morariu M. C. *Using a monopole sound power source to determine machinery sound power.* Euronoise Proceedings. 2018, pp. 599-606.

[Ver06] Vér, I. L. and Beranek, L. L. *Noise and vibration control engineering.* New Jersey: Wiley, 2006.

[Vor08] Vorländer M. *Auralization.* Berlin: Springer, 2008.

[Vor951] Vorländer M. and Raabe M. *Calibration of reference sound sources.* Acustica. 1995, Vol. 81(3), pp. 247-263.

[Vor952] Vorländer M. *Revised relation between the sound power and the average sound pressure levels in rooms and consequences for acoustic measurements.* Acustica. 1995, Vol. 81(4), pp. 332-343.

[Wit02] Wittstock V. and Bethke C. *The influence of temperature and static pressure on the sound power of aerodynamic reference sound sources.* Intenoise Proceedings. 2002, pp. 550-555.

[Wit041] Wittstock V. *On the uncertainty of meteorological corrections in sound power determination.* Internoise Proceedings. 2004, pp. 1675-1680.

[Wit042] Wittstock V. and Bethke C. *The influence of bandwidth on the qualification of anechoic and hemianechoic rooms.* Internoise Proceedings. 2004, pp. 1681-1686.

[Wit051] Wittstock V. and Bethke C. *On the uncertainty of sound pressure levels determined by third-octave band analyzers in a hemianechoic room.* Forum Acusticum Proceedings. 2005, pp. 1301-1304.

[Wit052] Wittstock V. *Uncertainties in building acoustics.* Forum Acusticum Proceedings. 2005, pp. 2217-2222.

[Wit13] Wittstock V., Schmelzer M. and Bethke C. *Establishing traceability for the quantity sound power.* Internoise Proceedings. 2013, pp. 3547-3551.

[Wit14] Wittstock V. *A new approach in sound power metrology.* OIML Bulletin. 2014, Vol. LV(2/3), pp. 9-13.

[Wu86] Wu M. Q. and Crocker M. J. *A computer simulated investigation of the estimation error of sound power measurement.* Internoise Proceedings. 1986, pp. 1129-1134.

[Yam14] Yamada K., Takahashi H. and Horiuchi R. *Influence of reflecting plane having finite surface density on sound power level of reference sound sources calibrated in hemi free-field.* Internoise Proceedings. 2014, pp. 550-558.

[Yam15] Yamada K., Takahashi H. and Horiuchi R. *Theoretical and experimental investigation of sound power transmitting through reflecting plane with low surface density in the calibration of reference sound sources*. Acoust. Sci. & Tech. 2015, Vol. 36(4), pp. 374-376.

[Yan82] Yanagisawa T. and Tsujita W. *Automated measuring system for sound power measurement*. Appl. Acoust. 1982, Vol. 15(6), pp. 445-457.

[Zho18] Zhong J., et al. *Effects on a finite size reflecting disk in sound power measurements*. Appl. Acoust. 2018, Vol. 140, pp. 24-29.

Appendix A: Supplement for chapter 2

Table A.1: Spherical coordinates of the position of original and mirror sources.

Original source	Mirror source
Floor reflection - Vertical translation	
Monopole	
$S=[z_0, 0, 0]$	$S'=[z_0, \pi, 0]$
Dipole	
$S_+=[z_0+d/2, 0, 0]$	$S'_+=[z_0+d/2, \pi, 0]$
$S_-=[z_0-d/2, 0, 0]$	$S'_-=[z_0-d/2, \pi, 0]$
Floor reflection - Horizontal translation	
Monopole	
$S=\left[\sqrt{x_0^2+z_0^2}, \tan^{-1}\left(\frac{\lvert x_0\rvert}{\lvert z_0\rvert}\right), \pi\right]$	
$S'=\left[\sqrt{x_0^2+z_0^2}, \pi-\tan^{-1}\left(\frac{\lvert x_0\rvert}{\lvert z_0\rvert}\right), \pi\right]$	
Dipole	
$S_+=\left[\sqrt{x_0^2+(z_0+d/2)^2}, \tan^{-1}\left(\frac{\lvert x_0\rvert}{\lvert z_0+d/2\rvert}\right), \pi\right]$	
$S_-=\left[\sqrt{x_0^2+(z_0-d/2)^2}, \tan^{-1}\left(\frac{\lvert x_0\rvert}{\lvert z_0-d/2\rvert}\right), \pi\right]$	
$S'_+=\left[\sqrt{x_0^2+(z_0+d/2)^2}, \pi-\tan^{-1}\left(\frac{\lvert x_0\rvert}{\lvert z_0+d/2\rvert}\right), \pi\right]$	
$S'_-=\left[\sqrt{x_0^2+(z_0-d/2)^2}, \pi-\tan^{-1}\left(\frac{\lvert x_0\rvert}{\lvert z_0-d/2\rvert}\right), \pi\right]$	
Side wall reflection - Vertical translation	
Monopole	
$S=[z_0, 0, 0]$	$S'=\left[\sqrt{(2\Delta x)^2+z_0^2}, \tan^{-1}\left(\frac{2\Delta x}{\lvert z_0\rvert}\right), 0\right]$

Dipole
$S'_{+} = [z_0 + d/2, \pi, 0]$
$S'_{-} = [z_0 - d/2, \pi, 0]$
$S'_{+} = \left[\sqrt{(2\Delta x)^2 + (z_0 + d/2)^2}, \tan^{-1}\left(\frac{2\Delta x}{\lvert z_0 + d/2 \rvert}\right), 0\right]$
$S'_{-} = \left[\sqrt{(2\Delta x)^2 + (z_0 - d/2)^2}, \tan^{-1}\left(\frac{2\Delta x}{\lvert z_0 - d/2 \rvert}\right), 0\right]$
Side wall reflection - Horizontal translation
Monopole
$S = \left[\sqrt{x_0^2 + z_0^2}, \tan^{-1}\left(\frac{\lvert x_0 \rvert}{\lvert z_0 \rvert}\right), \pi\right]$
$S' = \left[\sqrt{(2\Delta x - x_0)^2 + z_0^2}, \tan^{-1}\left(\frac{2\Delta x - x_0}{\lvert z_0 \rvert}\right), 0\right]$
Dipole
$S_{+} = \left[\sqrt{x_0^2 + (z_0 + d/2)^2}, \tan^{-1}\left(\frac{\lvert x_0 \rvert}{\lvert z_0 + d/2 \rvert}\right), \pi\right]$
$S_{-} = \left[\sqrt{x_0^2 + (z_0 - d/2)^2}, \tan^{-1}\left(\frac{\lvert x_0 \rvert}{\lvert z_0 - d/2 \rvert}\right), \pi\right]$
$S'_{+} = \left[\sqrt{(2\Delta x - x_0)^2 + (z_0 + d/2)^2}, \tan^{-1}\left(\frac{2\Delta x - x_0}{\lvert z_0 + d/2 \rvert}\right), 0\right]$
$S'_{-} = \left[\sqrt{(2\Delta x - x_0)^2 + (z_0 - d/2)^2}, \tan^{-1}\left(\frac{2\Delta x - x_0}{\lvert z_0 - d/2 \rvert}\right), 0\right]$

Table A.2: Specular reflection angle for each considered configuration.

Floor reflection - Vertical translation	
Monopole	$\theta_{S'} = \tan^{-1}\left(\dfrac{\sin\theta_{el}}{\cos\theta_{el} + L_z}\right)$
Dipole	$\theta_{S'_+} = \tan^{-1}\left(\dfrac{\sin\theta_{el}}{\cos\theta_{el} + L_z + D/2}\right)$
	$\theta_{S'_-} = \tan^{-1}\left(\dfrac{\sin\theta_{el}}{\cos\theta_{el} + L_z - D/2}\right)$
Floor reflection - Horizontal translation	
Monopole	$\theta_{S'} = \tan^{-1}\left(\dfrac{\sqrt{L_x^2 - 2L_x\sin\theta_{el}\cos\varphi_{az} + \sin^2\theta_{el}}}{\cos\theta_{el} + L_z}\right)$
Dipole	$\theta_{S'_+} = \tan^{-1}\left(\dfrac{\sqrt{L_x^2 - 2L_x\sin\theta_{el}\cos\varphi_{az} + \sin^2\theta_{el}}}{\cos\theta_{el} + L_z + D/2}\right)$
	$\theta_{S'_-} = \tan^{-1}\left(\dfrac{\sqrt{L_x^2 - 2L_x\sin\theta_{el}\cos\varphi_{az} + \sin^2\theta_{el}}}{\cos\theta_{el} + L_z - D/2}\right)$
Side wall reflection - Vertical translation	
Monopole	
$\theta_{S'} = \tan^{-1}\left(\dfrac{\sqrt{L_z^2 - 2L_z\cos\theta_{el} + \cos^2\theta_{el} + \sin^2\theta_{el}\sin^2\varphi_{az}}}{2\Delta x - \sin\theta_{el}\cos\varphi_{az}}\right)$	
Dipole	
$\theta_{S'_+} = \tan^{-1}\left(\dfrac{\sqrt{(L_z + D/2)^2 - 2(L_z + D/2)\cos\theta_{el} + \cos^2\theta_{el} + \sin^2\theta_{el}\sin^2\varphi_{az}}}{2\Delta x - \sin\theta_{el}\cos\varphi_{az}}\right)$	
$\theta_{S'_-} = \tan^{-1}\left(\dfrac{\sqrt{(L_z - D/2)^2 - 2(L_z - D/2)\cos\theta_{el} + \cos^2\theta_{el} + \sin^2\theta_{el}\sin^2\varphi_{az}}}{2\Delta x - \sin\theta_{el}\cos\varphi_{az}}\right)$	

Side wall reflection - Horizontal translation
Monopole
$\theta_{S'} = \tan^{-1}\left(\frac{\sqrt{L_z^2 - 2L_z\cos\theta_{el} + \cos^2\theta_{el} + \sin^2\theta_{el}\sin^2\varphi_{az}}}{(2\Delta x - L_x) - \sin\theta_{el}\cos\varphi_{az}}\right)$
Dipole
$\theta_{S'_+} = \tan^{-1}\left(\frac{\sqrt{(L_z + D/2)^2 - 2(L_z + D/2)\cos\theta_{el} + \cos^2\theta_{el} + \sin^2\theta_{el}\sin^2\varphi_{az}}}{(2\Delta x - L_x) - \sin\theta_{el}\cos\varphi_{az}}\right)$
$\theta_{S'_-} = \tan^{-1}\left(\frac{\sqrt{(L_z - D/2)^2 - 2(L_z - D/2)\cos\theta_{el} + \cos^2\theta_{el} + \sin^2\theta_{el}\sin^2\varphi_{az}}}{(2\Delta x - L_x) - \sin\theta_{el}\cos\varphi_{az}}\right)$

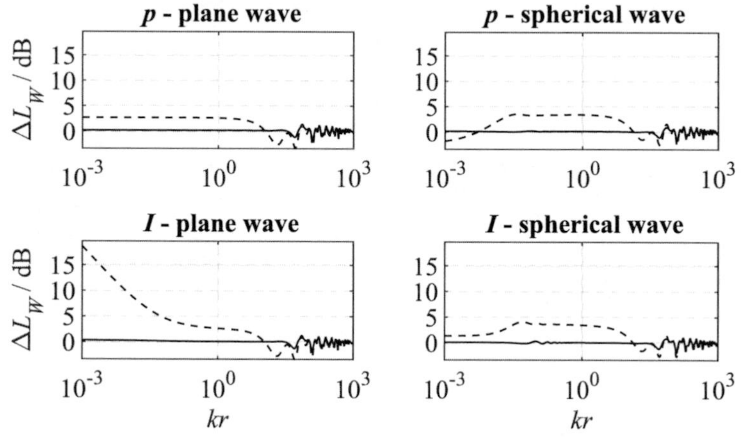

Figure A.1: Deviation from the free field sound power level after applying the direct (dashed) and the substitution method (continuous). Known source: monopole. Unknown source: monopole. Horizontal translation over reflecting floor. $L_x = 0.3$.

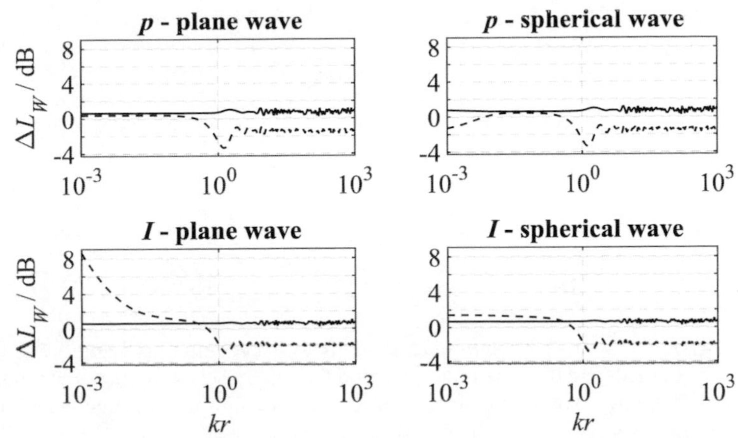

Figure A.2: Deviation from the free field sound power level after applying the direct (dashed) and the substitution method (continuous). Known source: monopole. Unknown source: monopole. Vertical translation next to absorbing side wall. $L_z = 0.3$.

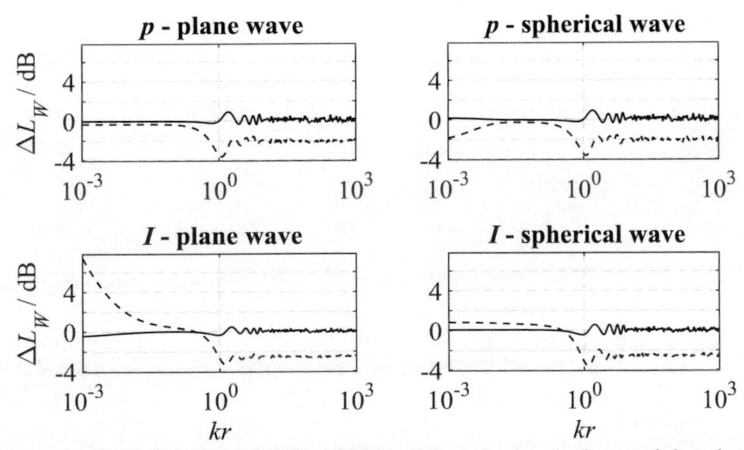

Figure A.3: Deviation from the free field sound power level after applying the direct (dashed) and the substitution method (continuous). Known source: monopole. Unknown source: monopole. Horizontal translation next to absorbing side wall. $L_x = 0.3$.

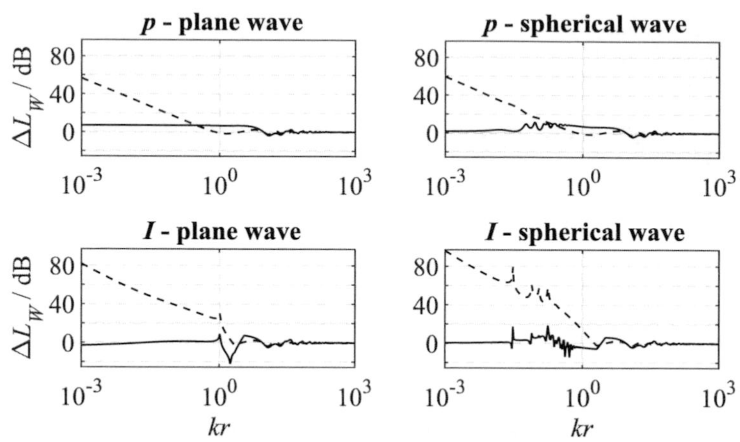

Figure A.4: Deviation from the free field sound power level after applying the direct (dashed) and the substitution method (continuous). Known source: dipole. Unknown source: dipole. Vertical translation over reflecting floor. $L_z = 0.3$.

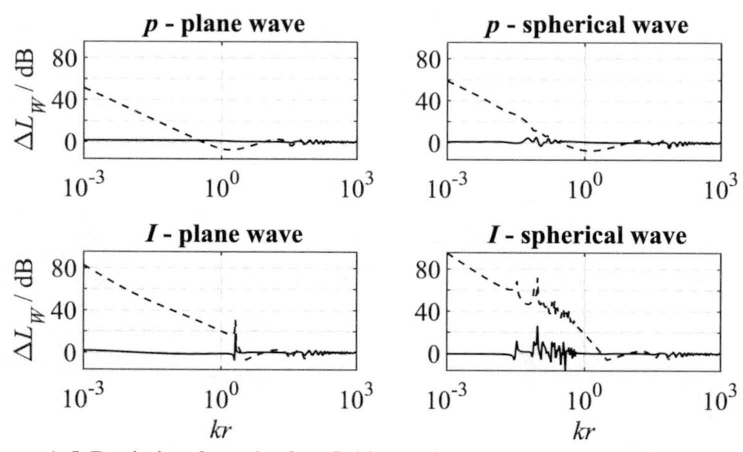

Figure A.5: Deviation from the free field sound power level after applying the direct (dashed) and the substitution method (continuous). Known source: dipole. Unknown source: dipole. Horizontal translation over reflecting floor. $L_x = 0.3$.

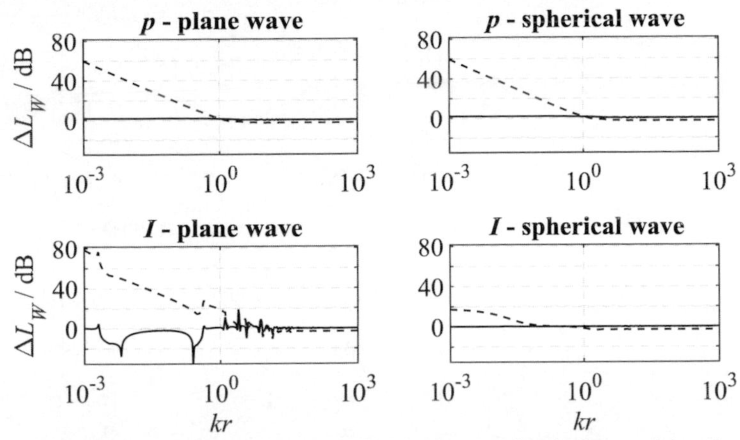

Figure A.6: Deviation from the free field sound power level after applying the direct (dashed) and the substitution method (continuous). Known source: dipole. Unknown source: dipole. Vertical translation next to absorbing side wall. $L_z = 0.3$.

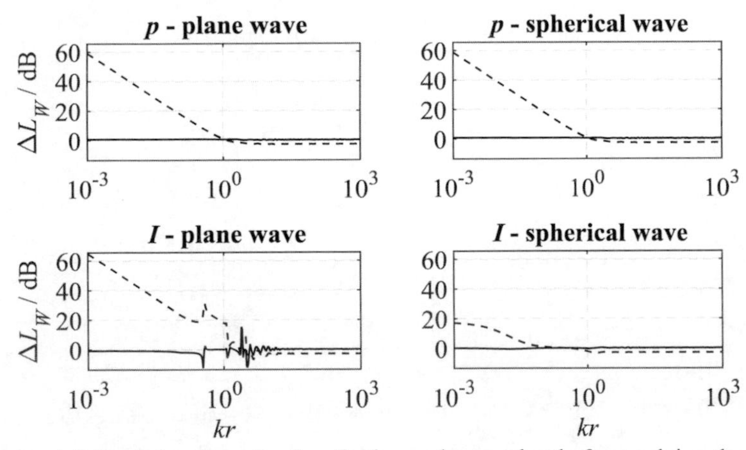

Figure A.7: Deviation from the free field sound power level after applying the direct (dashed) and the substitution method (continuous). Known source: dipole. Unknown source: dipole. Horizontal translation next to absorbing side wall. $L_x = 0.3$.

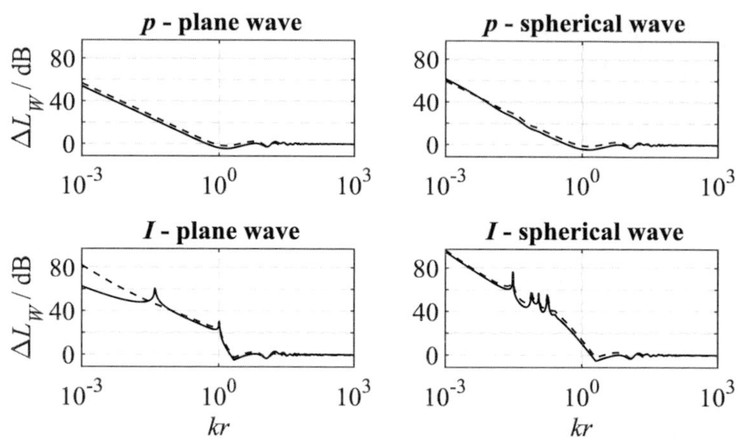

Figure A.8: Deviation from the free field sound power level after applying the direct (dashed) and the substitution method (continuous). Known source: monopole. Unknown source: dipole. Vertical translation over reflecting floor. $L_z = 0.3$.

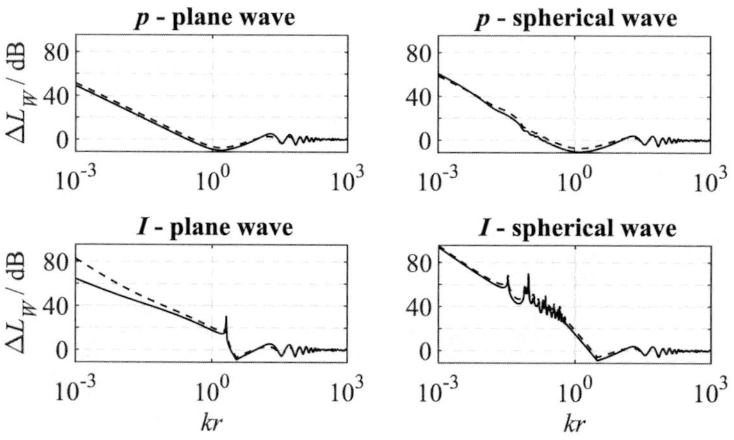

Figure A.9: Deviation from the free field sound power level after applying the direct (dashed) and the substitution method (continuous). Known source: monopole. Unknown source: dipole. Horizontal translation over reflecting floor. $L_x = 0.3$.

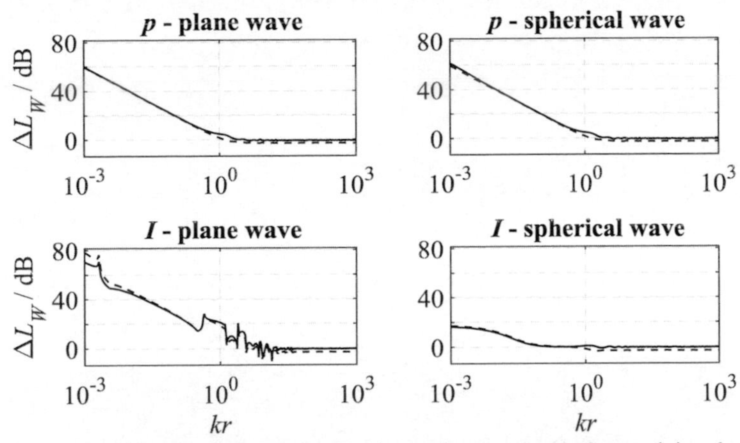

Figure A.10: Deviation from the free field sound power level after applying the direct (dashed) and the substitution method (continuous). Known source: monopole. Unknown source: dipole. Vertical translation next to absorbing side wall. $L_z = 0.3$.

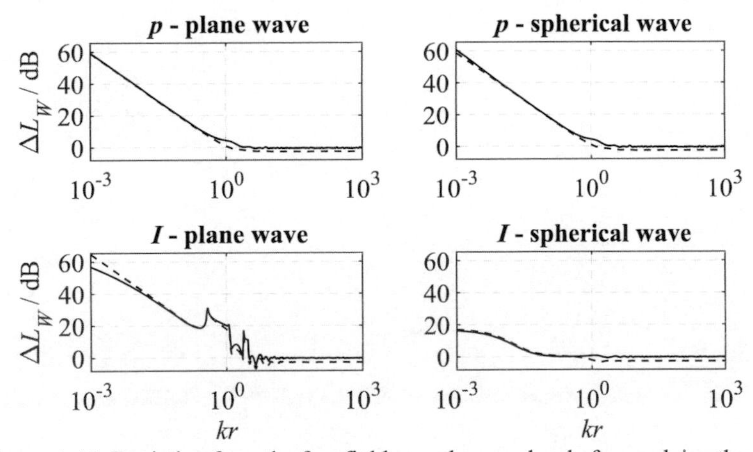

Figure A.11: Deviation from the free field sound power level after applying the direct (dashed) and the substitution method (continuous). Known source: monopole. Unknown source: dipole. Horizontal translation next to absorbing side wall. $L_x = 0.3$.

Appendix B: Supplement for chapter 3

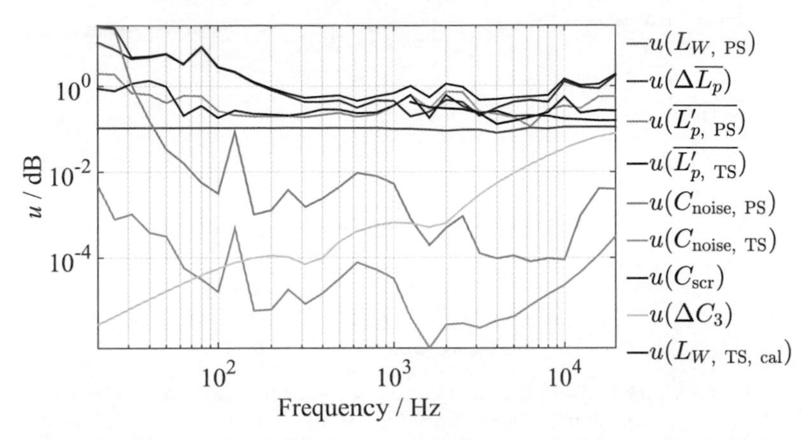

Figure B.1: Combined and individual uncertainty for the sound power determination under calibration conditions for one-third octave bands. B&K source.

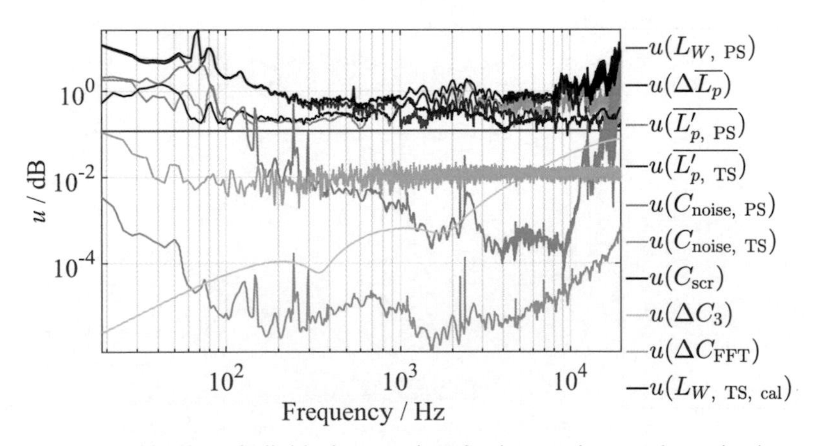

Figure B.2: Combined and individual uncertainty for the sound power determination under calibration conditions for FFT bands. B&K source.

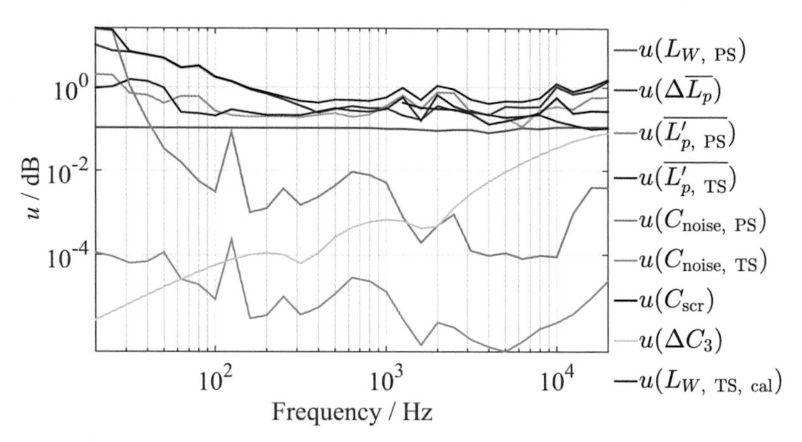

Figure B.3: Combined and individual uncertainty for the sound power determination under calibration conditions for one-third octave bands. EDF source.

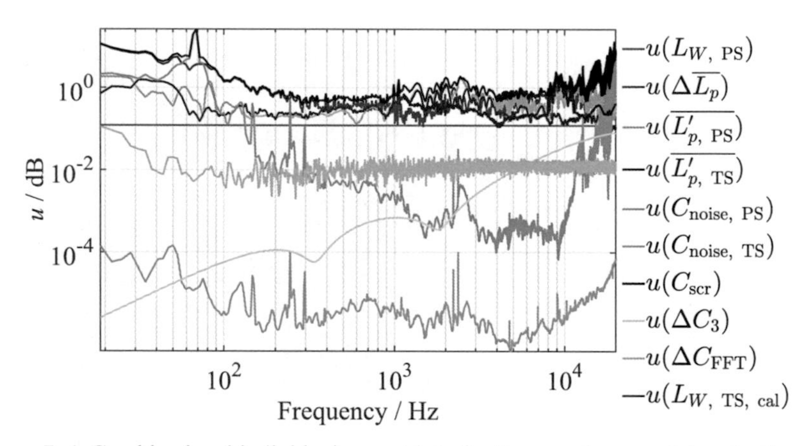

Figure B.4: Combined and individual uncertainty for the sound power determination under calibration conditions for FFT bands. EDF source.

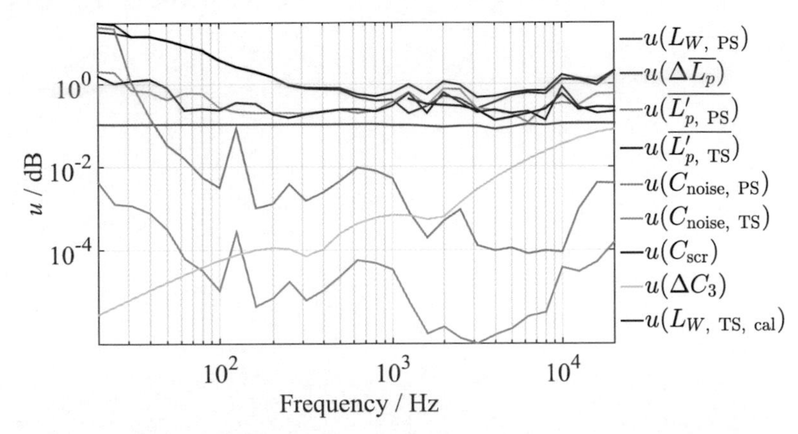

Figure B.5: Combined and individual uncertainty for the sound power determination under calibration conditions for one-third octave bands. NOR source.

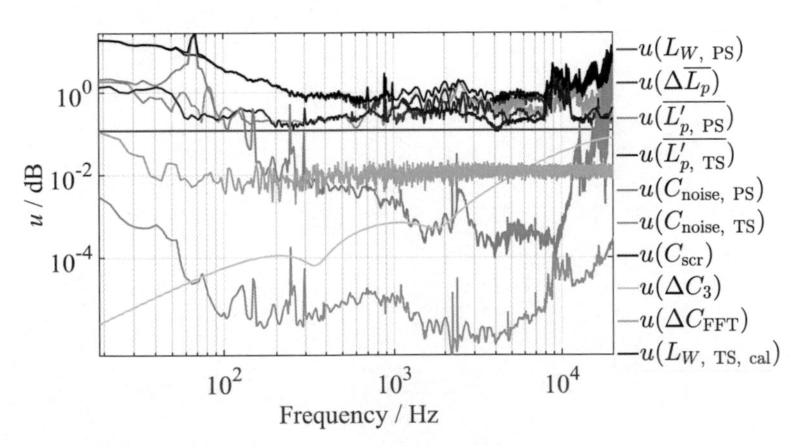

Figure B.6: Combined and individual uncertainty for the sound power determination under calibration conditions for FFT bands. NOR source.

Appendix C: Supplement for chapter 4

Table C.1: Environmental and operational parameters values for the determination of the sound power correction factor related to changes in atmospheric pressure.

AEG	B&K 1	B&K 2	B&K 3	EDF	NOR
$B_{\text{in situ}}$ / kPa					
99,06	97,52	99,17	99,26	99,31	99,39
99,27	99,62	99,62	99,65	99,69	99,45
99,5	99,64	99,68	99,85	100,1	99,63
100,15	99,65	99,71	99,89	100,15	99,71
101,01	99,86	99,73	100,08	100,23	99,8
101,05	99,94	100,24	100,22	100,96	101,05
101,06	101,53	101,53	100,25	101,16	101,51
101,1	102,03	102,08	101,15	101,23	102,07
101,53	102,1	102,13	101,51	101,52	102,2
101,55	102,16	102,15	101,63	101,91	102,21
99,06	97,52	99,17	99,26	99,31	99,39
B_{cal} /kPa					
101,98	102,48	102,23	102,23	102,51	102,47
$T_{\text{in situ}}$ / K					
291,6	291,3	292,1	292,2	291,6	292,6
292,6	292,4	292,35	292,5	292,15	292,7
292,4	292,7	292,5	293,15	292	292,65
292,55	292,3	293,2	292,55	291,6	291,8
292,4	292,6	292,45	292,25	292,15	292,3
291,8	292,75	291,7	292,25	293,4	292,6
292,2	292,25	292,7	292,95	292,75	293,3
293,05	291,6	292	291,5	293,55	291,45
291,95	292	292	292,85	292,9	291,55
292,05	292,15	292,05	291,3	292,25	292,35
291,6	291,3	292,1	292,2	291,6	292,6
T_{cal} /K					
293	293,2	293,2	292,9	292,25	293,3
$\omega_{\text{in situ}}$ / Hz					
47,53	48,21	48,08	47,99	44,3	49,09
47,8	48,24	48,14	48,05	44,23	48,99
47,9	48,27	48,12	48,03	44,82	49,08
47,4	48,3	48,13	48,03	45,06	49,07
47,67	48,27	48,18	48,03	44,6	49
47,67	48,23	48,11	48,01	44,47	49,01
47,42	48,26	48,11	48,02	44,48	49,07
47,72	48,24	48,1	48,01	44,07	49,07
47,35	48,22	48,08	48,03	44,57	49,03
47,62	48,23	48,13	47,98	45,05	49,08
47,53	48,21	48,08	47,99	44,3	49,09
ω_{cal}/Hz					
47,02	48,23	48,05	48	44,22	49,04

Table C.2: Environmental and operational parameters values for the determination of the sound power correction factor related to changes in ambient temperature.

B&K	**EDF**	**NOR**
	$B_{\text{in situ}}$ **/ kPa**	
101,5	101,45	101,48
101,47	101,67	101,58
100,46	100,47	100,48
100,65	100,55	100,57
102,21	101,1	102,18
101,32		100,92
101,5	101,45	101,48
	B_{cal} **/kPa**	
101,23	101,03	101,18
	$T_{\text{in situ}}$ **/ K**	
274,15	274,15	274,15
276,75	276,75	276,75
282,15	282,15	282,15
286,15	286,15	286,15
289,25	292,75	289,25
292,75		292,75
306,55	306,55	306,55
	T_{cal} **/K**	
286,82	286,42	286,82
	$\omega_{\text{in situ}}$ **/ Hz**	
48,23	43,7	49,02
48,25	42,67	48,98
48,3	43,92	49,03
48,27	42,6	49,05
48,23	44,23	49,07
48,17		49,07
48,23	43,7	49,02
	ω_{cal}**/Hz**	
48,24	43,4	49,02

Table C.3: Environmental and operational parameters values for the determination of the sound power correction factor related to changes in rotation speed.

	Current	$B_{\text{in situ}}$ **/ kPa**						B_{cal} **/ kPa**
	50 Hz	100,8	100,8	100,8	100,8	100,8	100,8	100,8
	60 Hz	100	100	100	100	100,1		100,02
		$T_{\text{in situ}}$ **/ K**						T_{cal} **/ K**
	50 Hz	297,15	296,15	296,15	296,15	297,15	297,15	296,65
	60 Hz	297,15	297,15	297,15	297,15	297,15		297,15
B&K		$\omega_{\text{in situ}}$ **/ Hz**						ω_{cal} **/ Hz**
	50 Hz	44,13	47,63	47,87	48,08	48,28	48,35	47,39
	60 Hz	53,25	54,35	55,27	55,77	56,27		54,98
		V_{in} **/ V**						
	50 Hz	150	207	218,5	230	241,5	253	
	60 Hz	130,5	109,25	115	120,75	126,6		

Table C.4: Selected microphone positions for the directivity uncertainty calculation.

Microphone position in degrees	**Number of microphones**						
8	**8**	**10**	**12**	**16**	**18**	**20**	**24**
17	x		x	x	x	x	x
29		x			x	x	x
38			x	x		x	x
44	x	x		x	x		x
51			x		x	x	x
57		x		x		x	x
63	x		x	x	x	x	x
68					x	x	x
73		x	x	x	x	x	x
78	x			x	x		x
83		x	x			x	x
88				x	x	x	x
92				x	x	x	x
97		x	x			x	x
102	x			x	x		x
107		x	x	x	x	x	x
112					x	x	x
117	x		x	x	x	x	x
123		x		x		x	x
129			x		x	x	x
136	x	x		x	x		x
142			x	x		x	x
151		x			x	x	x
163	x		x	x	x	x	x

Table C.5: Range of time and surface averaged sound pressure level difference for different number of microphones.

	Sound pressure level difference range in dB					
	B&K		**EDF**		**NOR**	
	1/3 Oct	FFT	1/3 Oct	FFT	1/3 Oct	FFT
$\overline{L_{p,8}} - \overline{L_{p,24}}$	0,93	1,27	0,71	1,52	0,87	3,05
$\overline{L_{p,10}} - \overline{L_{p,24}}$	0,58	1,12	0,51	0,98	0,48	1,91
$\overline{L_{p,12}} - \overline{L_{p,24}}$	0,45	0,7	0,35	0,86	0,44	1,87
$\overline{L_{p,16}} - \overline{L_{p,24}}$	0,27	0,54	0,25	0,61	0,3	1,31
$\overline{L_{p,18}} - \overline{L_{p,24}}$	0,26	0,38	0,21	0,5	0,22	0,99
$\overline{L_{p,20}} - \overline{L_{p,24}}$	0,2	0,37	0,21	0,43	0,23	0,72

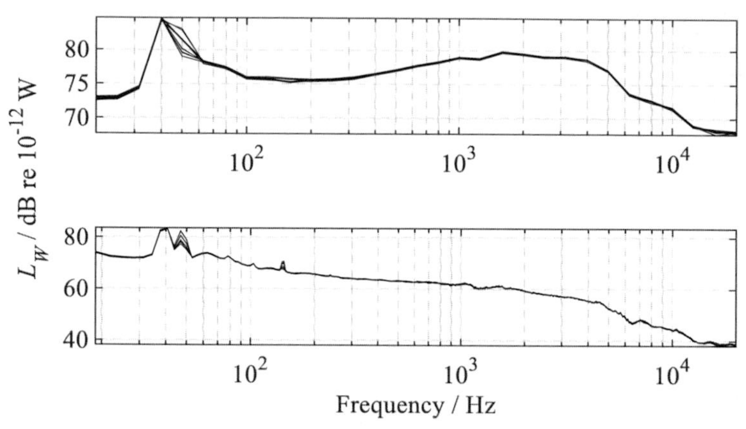

Figure C.1: Sound power level of the AEG source at various atmospheric pressures for one-third octave bands (top) and FFT bands (bottom).

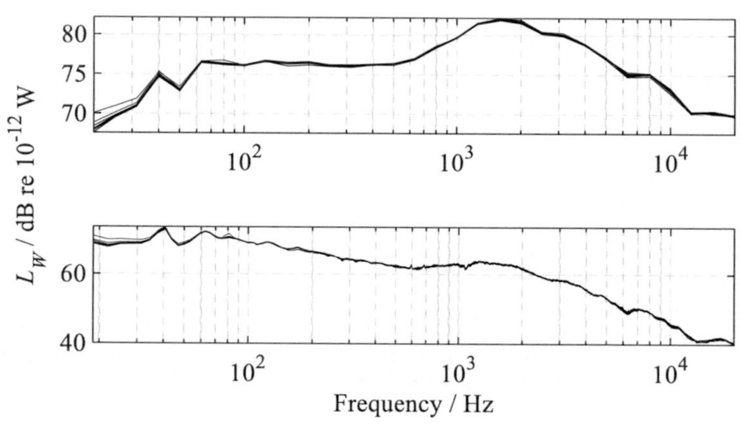

Figure C.2: Sound power level of the first B&K source at various atmospheric pressures for one-third octave bands (top) and FFT bands (bottom).

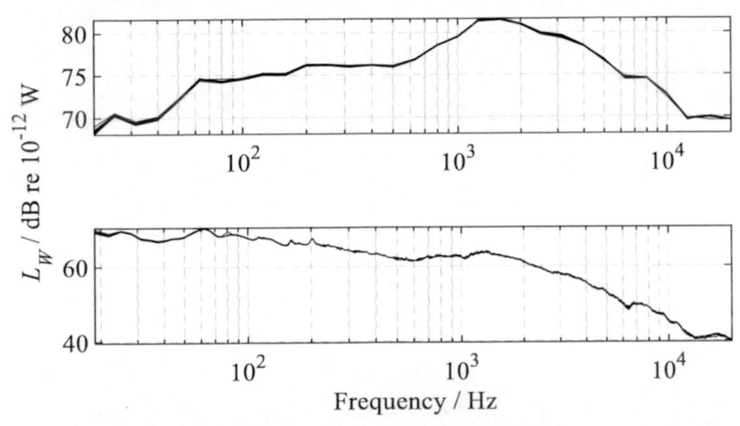

Figure C.3: Sound power level of the second B&K source at various atmospheric pressures for one-third octave bands (top) and FFT bands (bottom).

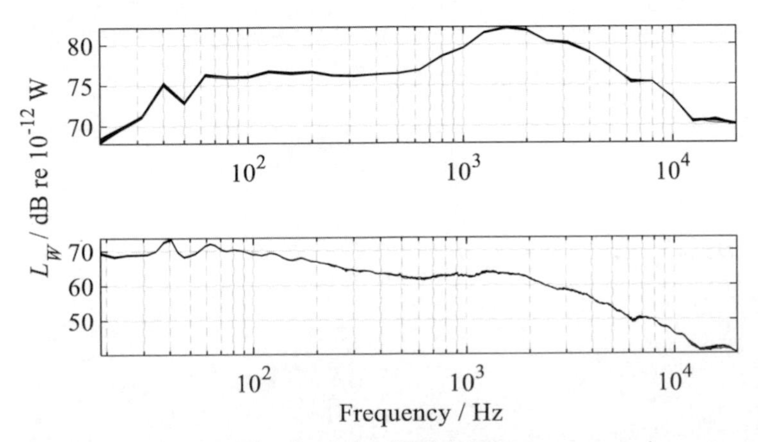

Figure C.4: Sound power level of the third B&K source at various atmospheric pressures for one-third octave bands (top) and FFT bands (bottom).

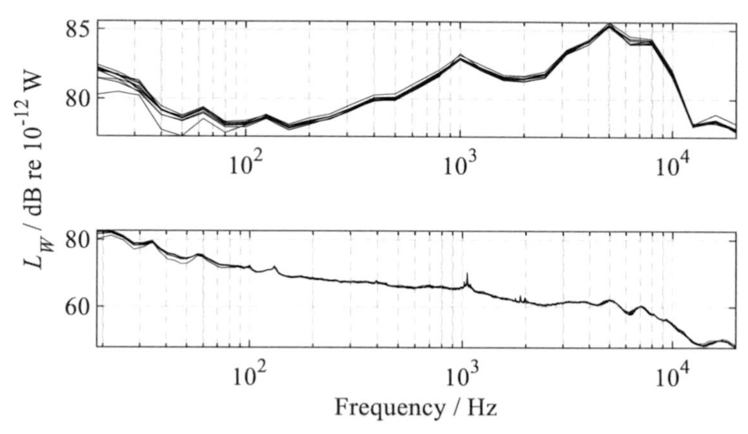

Figure C.5: Sound power level of the EDF source at various atmospheric pressures for one-third octave bands (top) and FFT bands (bottom).

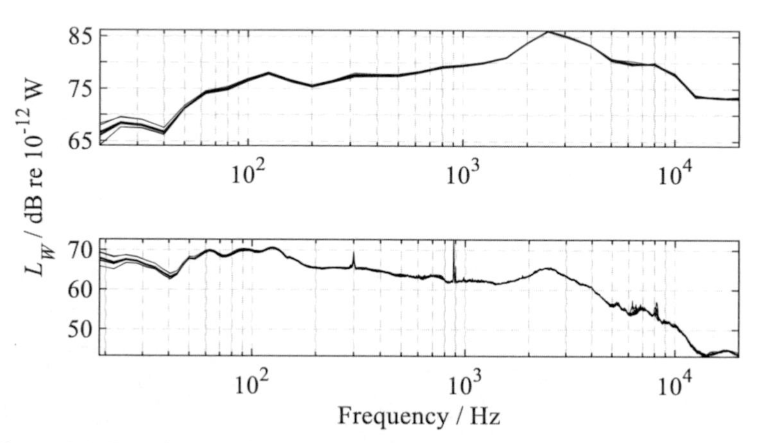

Figure C.6: Sound power level of the NOR source at various atmospheric pressures for one-third octave bands (top) and FFT bands (bottom).

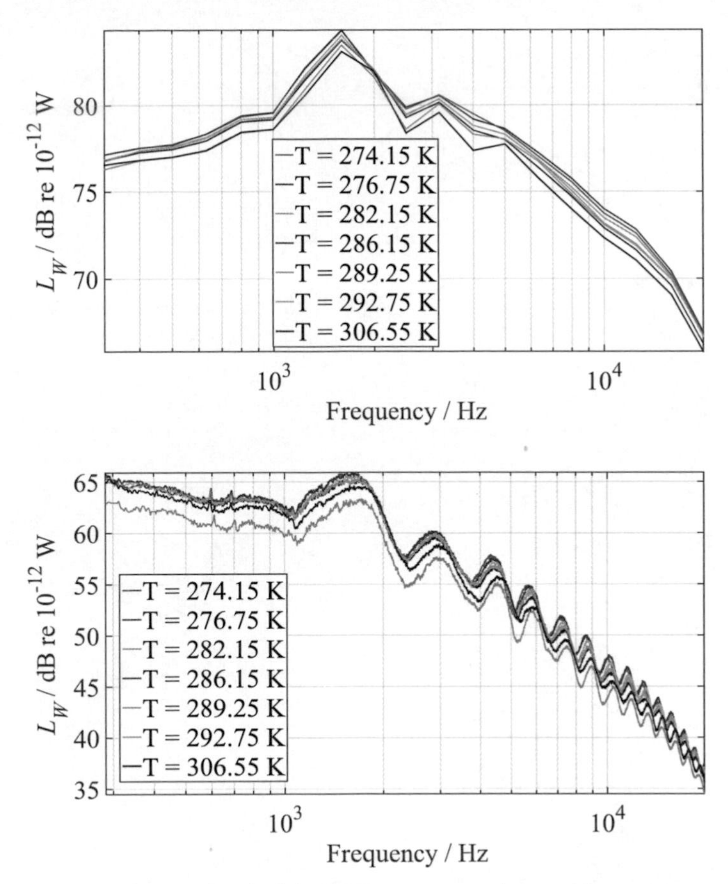

Figure C.7: Sound power level of the B&K source at various ambient temperatures for one-third octave bands (top) and FFT bands (bottom).

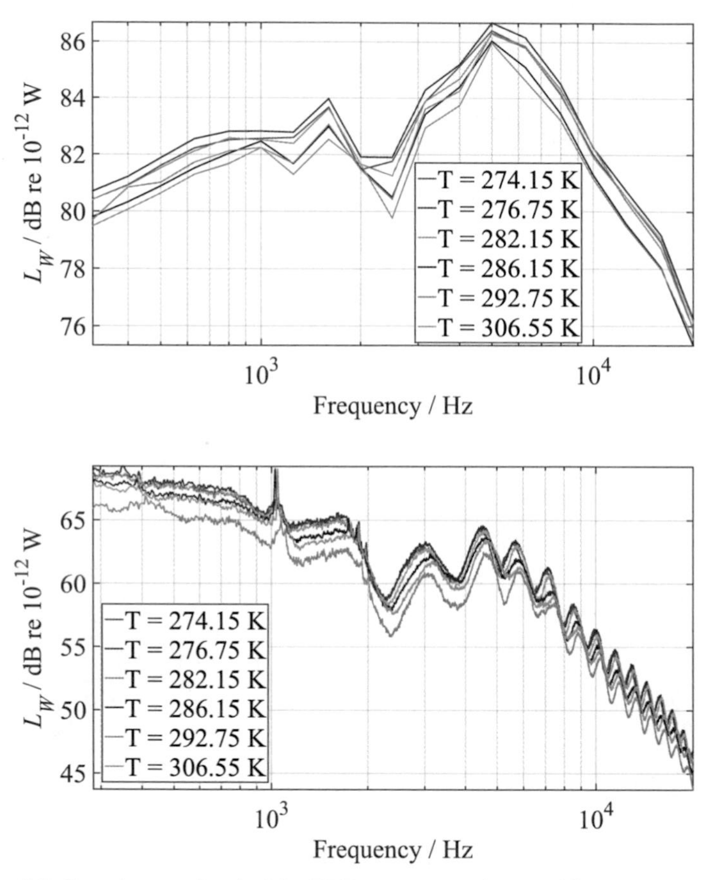

Figure C.8: Sound power level of the EDF source at various ambient temperatures for one-third octave bands (top) and FFT bands (bottom).

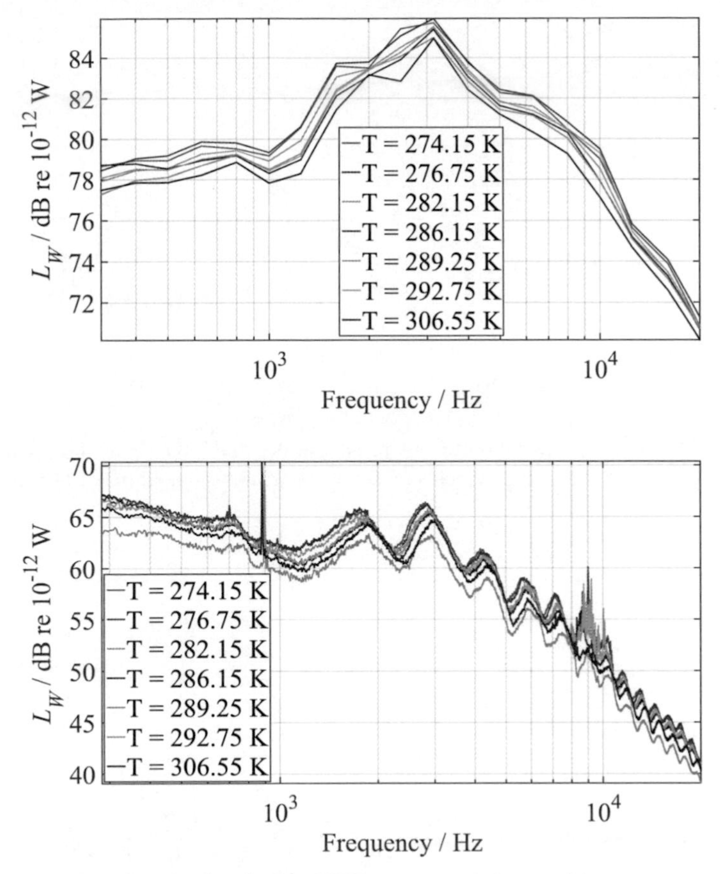

Figure C.9: Sound power level of the NOR source at various ambient temperatures for one-third octave bands (top) and FFT bands (bottom).

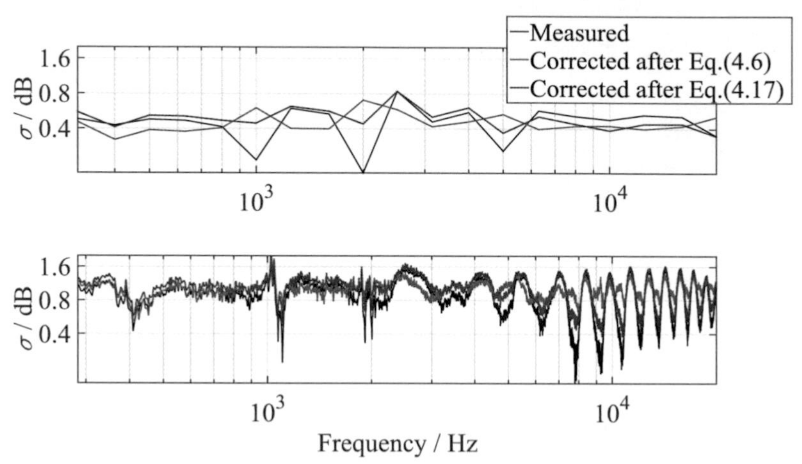

Figure C.10: Standard deviation of measured and corrected data after Eqs.(4.6) and (4.17) for measurements at various ambient temperatures. EDF source.

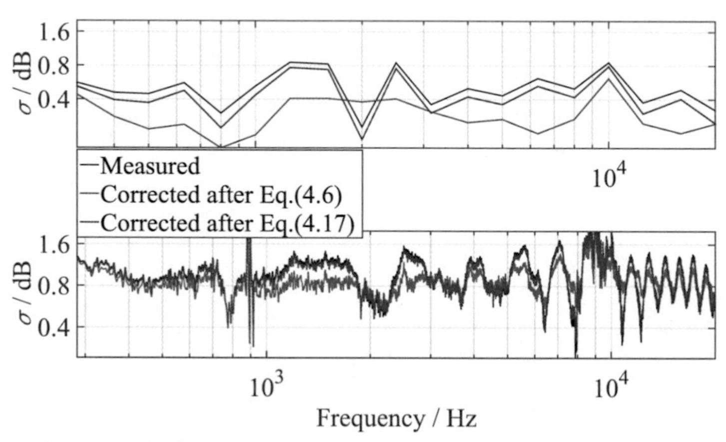

Figure C.11: Standard deviation of measured and corrected data after Eqs.(4.6) and (4.17) for measurements at various ambient temperatures. NOR source.

Appendix D: Supplement for chapter 5

Table D.1: Environmental conditions and measurement surface values for the investigation of the influence of the measurement surface.

Scanning measurements							
S/m^2	3,08	4,02	5,09	13,21	18,16	34,4	47,52
Primary source							
B_{cal}/kPa	101,07	101,04	101,01	100,37	100,36	100,36	100,25
T_{cal}/K	297	297,03	297,03	297,73	297,68	297,7	297,78
Transfer source							
B_{cal}/kPa	101,4	101,42	101,42	100,75	100,76	100,8	100,67
T_{cal}/K	297,6	297,68	297,5	297,85	297,9	297,88	297,13
AEG							
B_{cal}/kPa				100,48			
T_{cal}/K				298,13			
Air compressor							
B_{cal}/kPa				100,55			
T_{cal}/K				299,3			
Vacuum cleaner							
B_{cal}/kPa				100,59			
T_{cal}/K				298,83			
Box measurements							
S/m^2							
21,25							
AEG							
B_{cal}/kPa				101			
T_{cal}/K				297,15			
Air compressor							
B_{cal}/kPa				101			
T_{cal}/K				297,15			
Vacuum cleaner							
B_{cal}/kPa				100,7			
T_{cal}/K				296,15			

Table D.2: Environmental conditions, operational conditions and measurement surface values for the investigation of the influence of the surrounding environment.

Measurements under calibration conditions (scanning)							
S/m^2	3,08	4,02	5,09	13,21	18,16	34,4	47,52
Primary source							
B_{cal}/kPa	101,07	101,04	101,01	100,37	100,36	100,36	100,25
T_{cal}/K	297	297,03	297,03	297,73	297,68	297,7	297,78
Transfer source							
B_{cal}/kPa	101,4	101,42	101,42	100,75	100,76	100,8	100,67
T_{cal}/K	297,6	297,68	297,5	297,85	297,9	297,88	297,13
ω_{cal} / Hz	48,07	48,08	48,07	48,1	48,11	48,11	48,13
In situ measurements (box)							
S/m^2	21,25						
	Hard-walled 1	**Hard-walled 2**	**Hard-walled 2 (damped)**	**Hemianechoic**	**Open space**		
Transfer source							
B_{cal}/kPa	101,3	101,25	101,25	101	100,3		
T_{cal}/K	294,15	300,15	302,65	297,15	294,15		
ω_{cal} / Hz	48,12	48,12	48,12	48,12	48,12		
AEG							
B_{cal} / kPa	101,3	101,2	101,3	100,7	100,45		
T_{cal} /K	294,15	300,65	302,65	297,15	294,15		
Air compressor							
B_{cal} / kPa	101,15	101,25	101,25	100,7	100,45		
T_{cal} /K	294,65	301,65	302,65	296,15	294,15		
Vacuum cleaner							
B_{cal} / kPa	101,15	101,25	101,25	100,7	100,45		
T_{cal} /K	294,65	301,65	302,65	296,15	294,15		

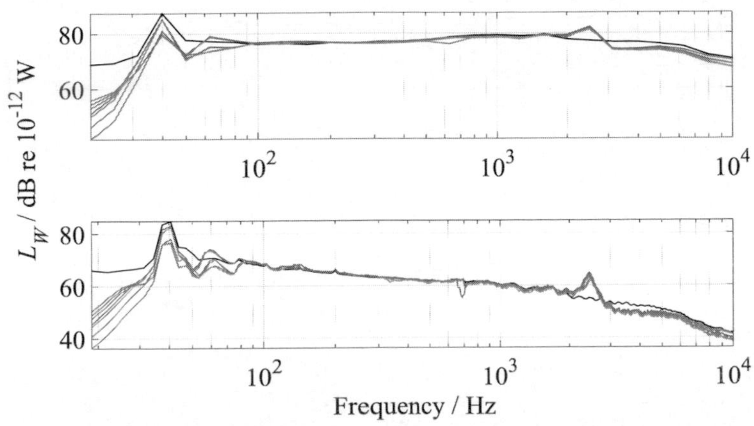

Figure D.1: Sound power level of the AEG source directly determined after sound pressure scanning measurements (black) and after applying the substitution method (grey). Top: one-third octave bands. Bottom: FFT bands.

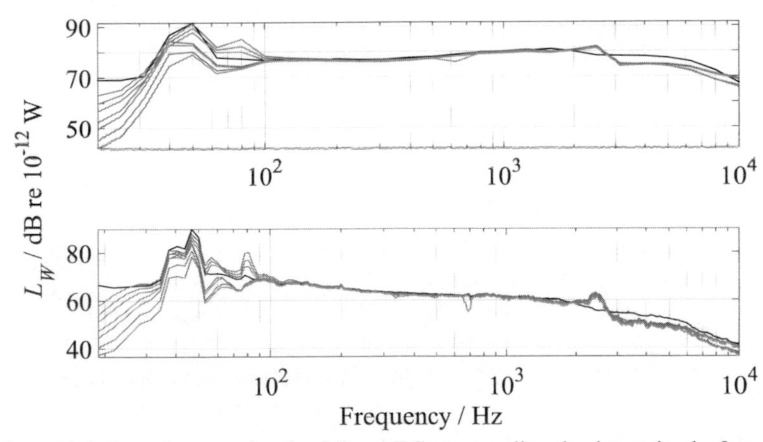

Figure D.2: Sound power level of the AEG source directly determined after sound pressure discrete point measurements (black) and after applying the substitution method (grey). Top: one-third octave bands. Bottom: FFT bands.

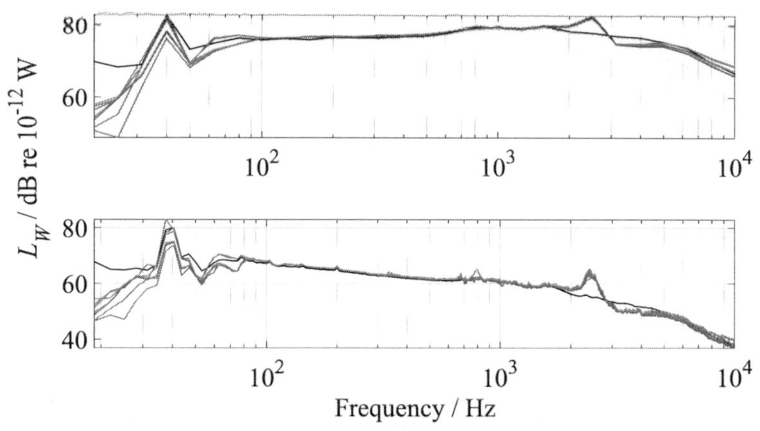

Figure D.3: Sound power level of the AEG source directly determined after sound intensity scanning measurements (black) and after applying the substitution method (grey). Top: one-third octave bands. Bottom: FFT bands.

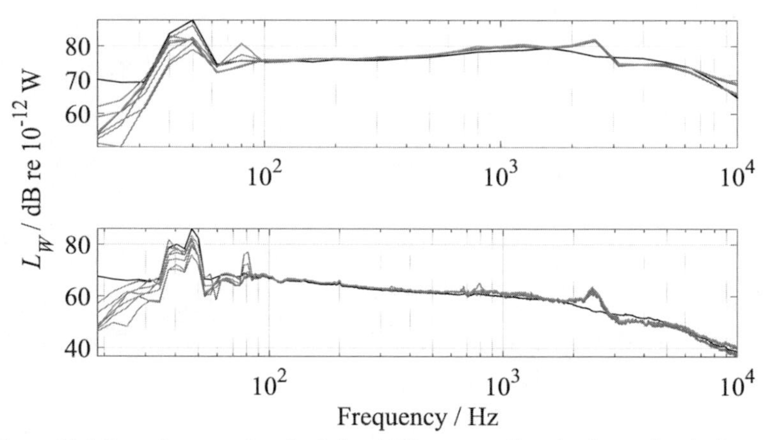

Figure D.4: Sound power level of the AEG source directly determined after sound intensity discrete point measurements (black) and after applying the substitution method (grey). Top: one-third octave bands. Bottom: FFT bands.

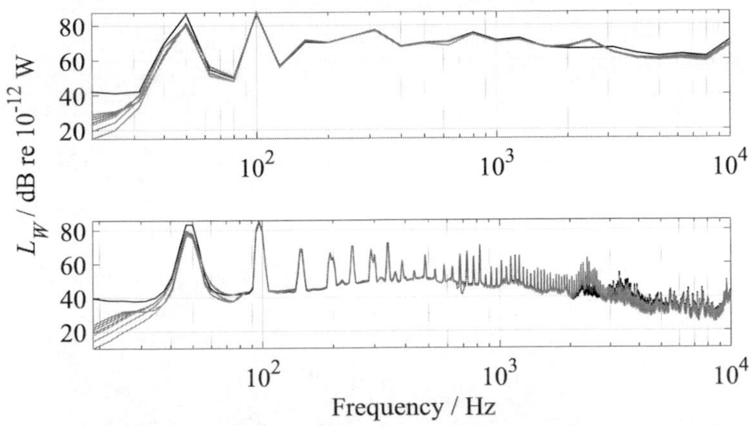

Figure D.5: Sound power level of the air compressor directly determined after sound pressure scanning measurements (black) and after applying the substitution method (grey). Top: one-third octave bands. Bottom: FFT bands.

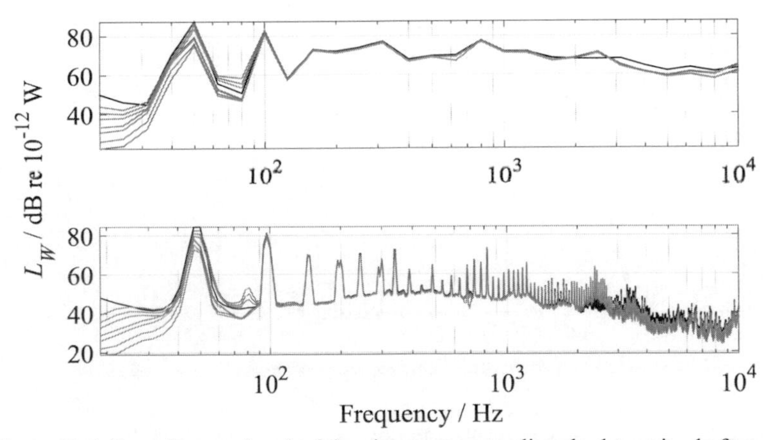

Figure D.6: Sound power level of the air compressor directly determined after sound pressure discrete point measurements (black) and after applying the substitution method (grey). Top: one-third octave bands. Bottom: FFT bands.

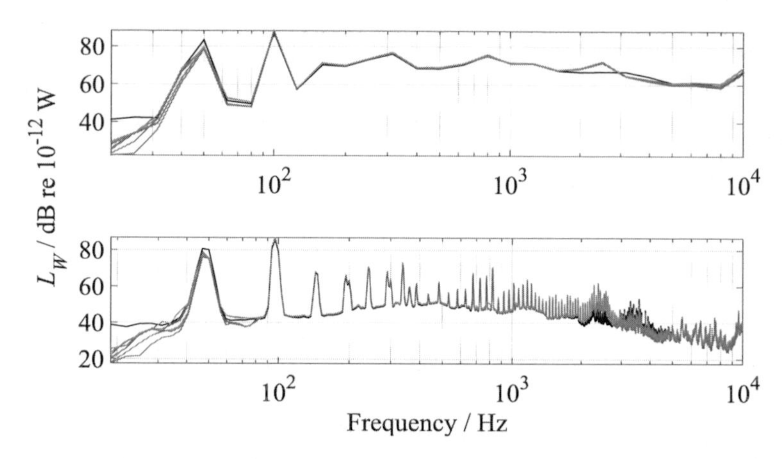

Figure D.7: Sound power level of the air compressor directly determined after sound intensity scanning measurements (black) and after applying the substitution method (grey). Top: one-third octave bands. Bottom: FFT bands.

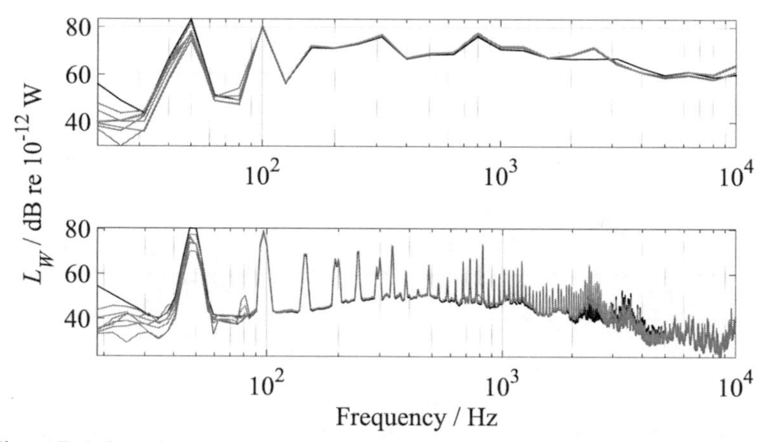

Figure D.8: Sound power level of the air compressor directly determined after sound intensity discrete point measurements (black) and after applying the substitution method (grey). Top: one-third octave bands. Bottom: FFT bands.

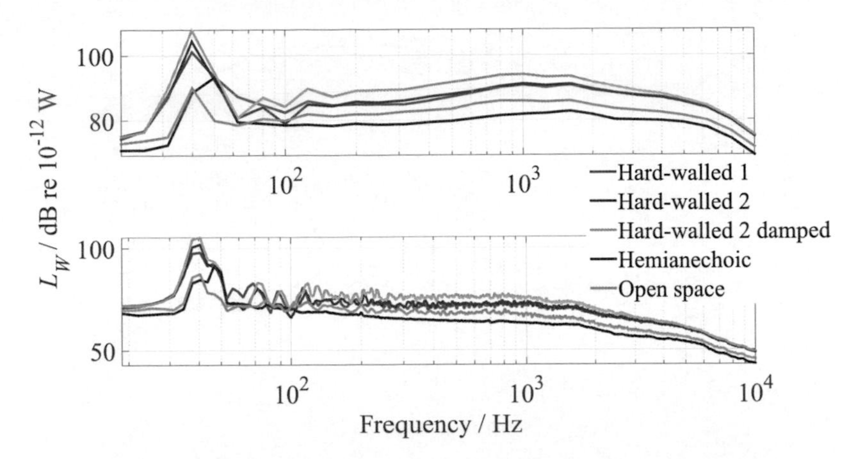

Figure D.9: Sound power level of the AEG source directly determined after sound pressure discrete point measurements at different acoustic environments. Top: one-third octave bands. Bottom: FFT bands.

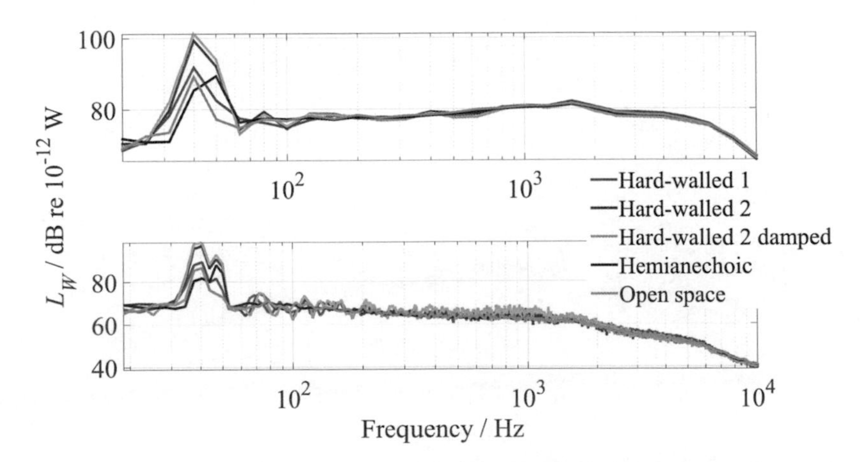

Figure D.10: Sound power level of the AEG source directly determined after sound intensity discrete point measurements at different acoustic environments. Top: one-third octave bands. Bottom: FFT bands.

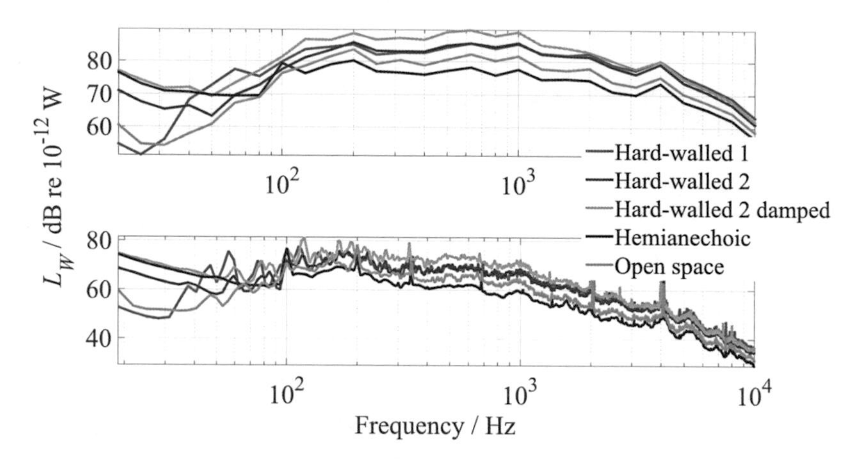

Figure D.11: Sound power level of the vacuum cleaner directly determined after sound pressure discrete point measurements at different acoustic environments. Top: one-third octave bands. Bottom: FFT bands.

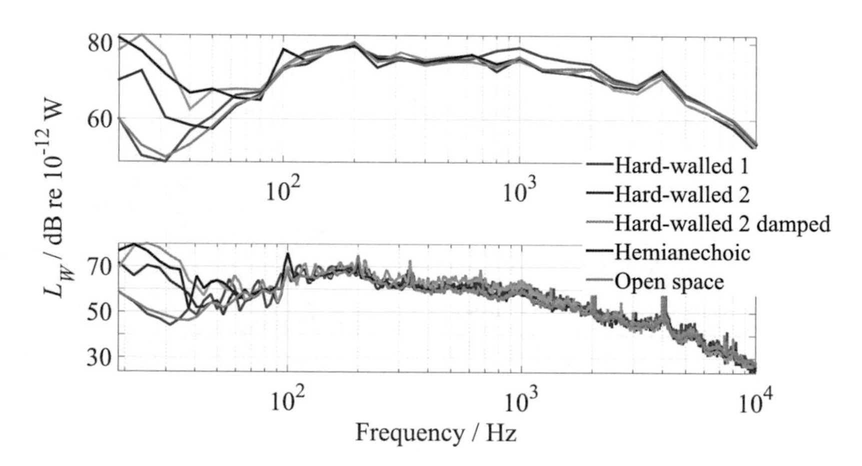

Figure D.12: Sound power level of the vacuum cleaner directly determined after sound intensity discrete point measurements at different acoustic environments. Top: one-third octave bands. Bottom: FFT bands.

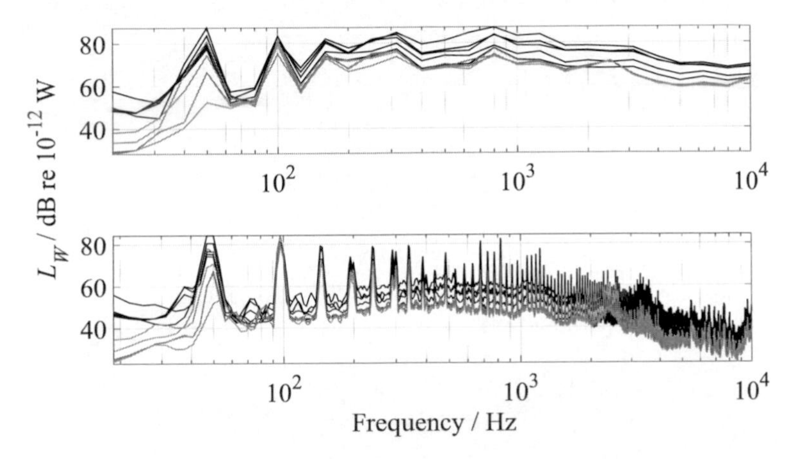

Figure D.13: Sound power level of the air compressor directly determined after sound pressure discrete point measurements (black) and after applying the substitution method (grey) at different surrounding environments. Top: one-third octave bands. Bottom: FFT bands.

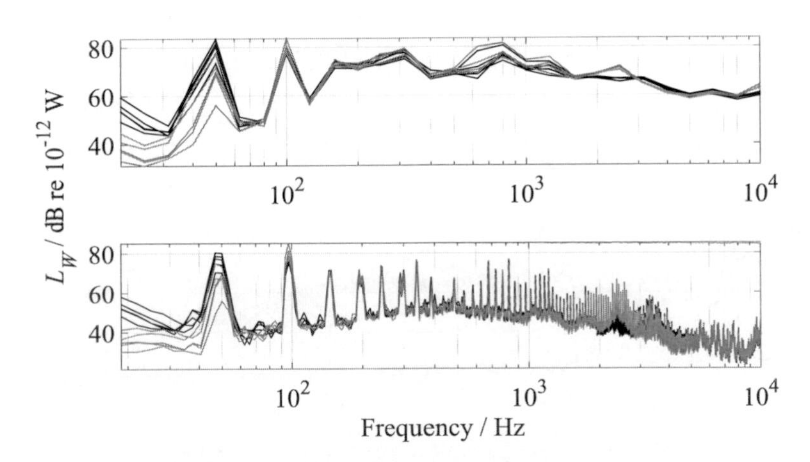

Figure D.14: Sound power level of the air compressor directly determined after sound intensity discrete point measurements (black) and after applying the substitution method (grey) at different surrounding environments. Top: one-third octave bands. Bottom: FFT bands.

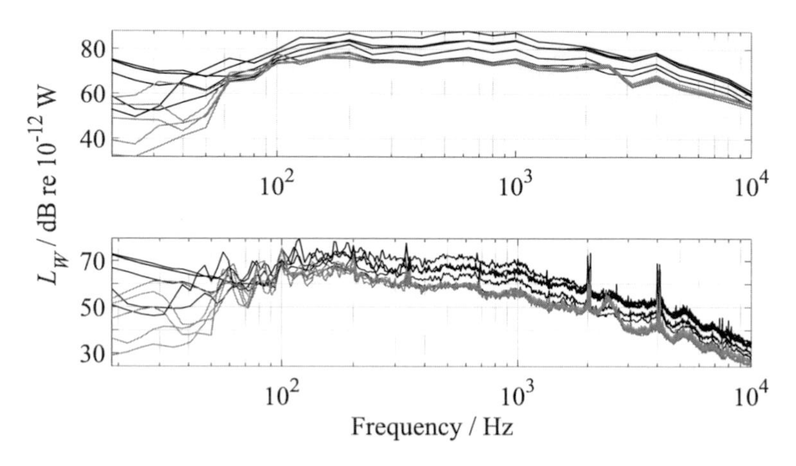

Figure D.15: Sound power level of the vacuum cleaner directly determined after sound pressure discrete point measurements (black) and after applying the substitution method (grey) at different surrounding environments. Top: one-third octave bands. Bottom: FFT bands.

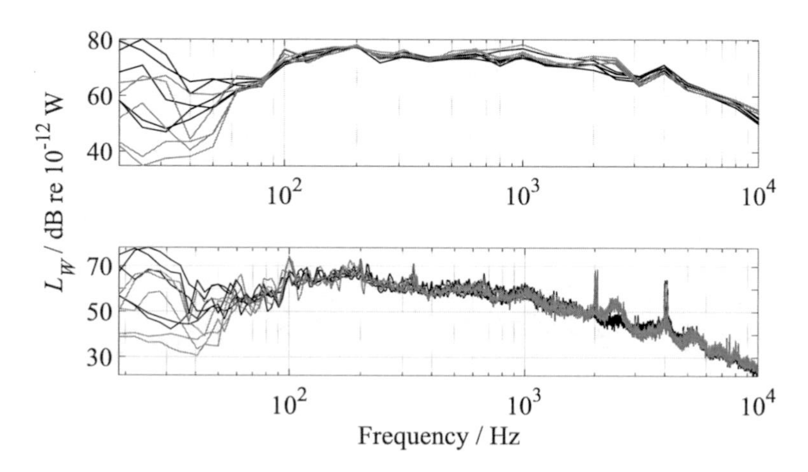

Figure D.16: Sound power level of the vacuum cleaner directly determined after sound intensity discrete point measurements (black) and after applying the substitution method (grey) at different surrounding environments. Top: one-third octave bands. Bottom: FFT bands.

List of figures

List of symbols

α	absorption coefficient
Γ_θ	spherical wave reflection angle
Δ	distance between two points
$\Delta\theta$	angle step for spherical wave reflection
Δx	distance between axis origin and side wall
Δt	time step
ΔL_p	sound pressure level difference between measured and after the spherical harmonics transform levels
$\Delta L_{p/I}$	sound pressure or sound intensity level difference
ΔL_W	sound power level difference
$\Delta L_W(B)$	sound power level difference between levels determined at different atmospheric pressure values
$\Delta L_{W,\mathrm{regression}}(B)$	sound power level difference after applying regression analysis to levels at different atmospheric pressure values
$\Delta L_{W,\mathrm{theo},i}$	theoretical near field error at *i*-th radius
$\Delta L_{W,\mathrm{meas},i}$	near field error based on measurements at *i*-th radius
Δr	sound intensity spacer length
$\Delta(r)$	theoretical near field error
Θ_1	real reflection angle for spherical waves
Θ_2	imaginary reflection angle for spherical waves
θ	reflection angle for spherical waves
θ_{el}	elevation angle
θ_{S_+}	reflection angle of the sound of the positive pole of the dipole source
θ_{S_-}	reflection angle of the sound of the negative pole of the dipole source
θ_0	reflection angle for plane waves
M	number of helices
ξ	specific acoustic impedance

π	Archimedes' constant
ρ	air density
$\sigma(\cdot)$	standard deviation of the quantity in parenthesis
σ_{r}	standard deviation under repeatability conditions
φ_{az}	azimuthal angle
ω	fan rotation speed
ω_{ang}	angular velocity
ω_{cal}	fan rotation speed under calibration conditions
$\omega_{\mathrm{in\ situ}}$	fan rotation speed in situ
$\omega_{\mathrm{nominal}}$	nominal fan rotation speed
B	atmospheric pressure
B_{band}	signal bandwidth
B_{cal}	atmospheric pressure under calibration conditions
$B_{\mathrm{DUT,\ cal}}$	atmospheric pressure during the measurement of a device under test under calibration conditions
$B_{\mathrm{DUT,\ in\ situ}}$	atmospheric pressure during the measurement of a device under test in situ
$B_{\mathrm{in\ situ}}$	atmospheric pressure in situ
B_{PS}	atmospheric pressure during the measurement of the primary source
$B_{\mathrm{TS,\ cal}}$	atmospheric pressure during the measurement of the transfer source under calibration conditions
$B_{\mathrm{TS,\ in\ situ}}$	atmospheric pressure during the measurement of the transfer source in situ
C	overall correction
$C_{\mathrm{ang},\,i}$	sound incidence correction
C_{att}	sound attenuation due to the scanning apparatus correction
$C_{\mathrm{cal},\,i}$	calibration values correction
C_{em}	sound emission correction
C_{FFT}	Hanning window correction

C_{fil}	filter correction
C_{nf}	near field correction
C_{noise}	background noise correction
$C_{\text{noise, DUT, cal}}$	background noise correction for a device under test under calibration conditions
$C_{\text{noise, PS}}$	background noise correction for the primary source
$C_{\text{noise, TS}}$	background noise correction for the transfer source
$C_{\text{noise, TS, cal}}$	background noise correction for the transfer source under calibration conditions
$C_{\text{pos}, i}$	position of the *i*-th microphone correction
C_{ref}	reflections from the scanning apparatus correction
C_{scr}	windscreen correction
$C_{\text{scr, probe}, p}$	sound intensity probe windscreen correction for sound pressure
$C_{\text{scr, probe}, I}$	sound intensity probe windscreen correction for sound intensity
C_1	reference quantity correction
C_2	acoustic radiation impedance correction
C_3	air absorption correction
$C_{3,\text{ DUT, cal}}$	air absorption correction for the measurement of a device under test under calibration conditions
$C_{3,\text{ PS}}$	air absorption correction for the measurement of the primary source
$C_{3,\text{ TS, cal}}$	air absorption correction for the measurement of the transfer source under calibration conditions
c	sound speed
∂	derivative
D	ratio of distance between dipole poles over radius
$D_{\text{I}i}$	directivity index of the *i*-th microphone
$D_{\text{I,max}}$	maximum directivity index
Dx	ratio of distance between axis origin and side wall over radius
d	distance between dipole poles

$d\Theta_1$	integration step for the real reflection angle of spherical waves
$d\Theta_2$	integration step for the imaginary reflection angle of spherical waves
$d\theta$	reflection angle integration step
$d\theta_{\mathrm{el}}$	elevation angle integration step
$d\varphi_{\mathrm{az}}$	azimuthal angle integration step
dS_i	surface area covered by the i-th receiver
dx	x integration step
E_n	energy of the n-th spherical harmonic
E_{tot}	total signal energy
$f(x)$	function of x
$f'(x)$	derivative of function of x
H	Helmholtz number
Ha	normalized Bessel function of the third kind
I	sound intensity
$\lvert I_{\mathrm{r}} \rvert$	unsigned magnitude of the radial (normal) component of the sound intensity
$I_{\mathrm{r}\,i}$	signed magnitude of the i-th partial surface averaged radial (normal) sound intensity
I_{r}	radial component of sound intensity vector
I_{ref}	sound intensity reference value
Im	imaginary part of complex number
J_0	Bessel function of zero order
j	complex number identity
k	wavenumber
$\overline{L_B}$	background noise level of all microphones
L_I	sound intensity level
$\overline{L_I}$	time and surface averaged sound intensity level

$\overline{L_{I,\text{dip, free field}}}$	free field time and surface averaged sound intensity level of dipole
$\overline{L_{I,\text{DUT, cal}}}$	time and surface averaged sound intensity level of a device under test under calibration conditions
$\overline{L_{I,\text{known}}}$	time and surface averaged sound intensity level of the known source
$\overline{L_{I,\text{mon, free field}}}$	free field time and surface averaged sound intensity level of monopole
$\overline{L_{I,\text{PS}}}$	time and surface averaged sound intensity level of the primary source
$\overline{L_{I,\text{TS, cal}}}$	time and surface averaged sound intensity level of the transfer source under calibration conditions
$\overline{L_{I,\text{unknown}}}$	time and surface averaged sound intensity level of the unknown source
$\overline{L_p}$	corrected time and surface averaged sound pressure level
$\overline{L'_p}$	uncorrected time and surface averaged sound pressure level
$L_{p,\text{arc}}$	sound pressure level measured with the scanning apparatus
$\overline{L_{p,\text{dip, free field}}}$	free field time and surface averaged sound pressure level of dipole
$\overline{L_{p,\text{DUT, cal}}}$	corrected time and surface averaged sound pressure level of a device under test under calibration conditions
$\overline{L'_{p,\text{DUT, cal}}}$	uncorrected time and surface averaged sound pressure level of a device under test under calibration conditions
$L'_{p,\text{ DUT, cal}, i}$	uncorrected sound pressure level of the *i*-th microphone for the measurement of a device under test under calibration conditions
$\overline{L_{p,\text{DUT, in situ}}}$	time and surface averaged sound pressure level of a device under test in situ
$\overline{L_{p,\text{known}}}$	time and surface averaged sound pressure level of the known source
$L_{p,\text{han},i}$	sound pressure level of the *i*-th microphone measured with Hanning window
$L_{p,\text{meas}}$	measured sound pressure level
$\overline{L_{p,\text{mon, free field}}}$	free field time and surface averaged sound pressure level of monopole
L_{pi}	sound pressure level of the *i*-th microphone

$\overline{L_{p/I,\,\mathrm{plane}}}$	time and surface averaged sound pressure or intensity level for plane wave
$\overline{L_{p/I,\,\mathrm{spherical}}}$	time and surface averaged sound pressure or intensity level for spherical wave
$\overline{L_{p,\,\mathrm{PS}}}$	corrected time and surface averaged sound pressure level of the primary source
$\overline{L'_{p,\,\mathrm{PS}}}$	uncorrected time and surface averaged sound pressure level of the primary source
$L_{p,\,\mathrm{PS},\,i}$	corrected sound pressure level of the *i*-th microphone for the measurement of the primary source
$L'_{p,\,\mathrm{PS},\,i}$	uncorrected sound pressure level of the *i*-th microphone for the measurement of the primary source
$L_{p,\,\mathrm{SHT}}$	sound pressure level after the spherical harmonics transform
$L_{p,\,\mathrm{stand}}$	sound pressure level measured with microphone stand
$\overline{L_{p,\,\mathrm{TS}}}$	corrected time and surface averaged sound pressure level of the transfer source
$\overline{L'_{p,\,\mathrm{TS}}}$	uncorrected time and surface averaged sound pressure level of the transfer source
$\overline{L_{p,\,\mathrm{TS,\,cal}}}$	corrected time and surface averaged sound pressure level of the transfer source under calibration conditions
$\overline{L'_{p,\,\mathrm{TS,\,cal}}}$	uncorrected time and surface averaged sound pressure level of the transfer source under calibration conditions
$L_{p,\,\mathrm{TS},\,i}$	corrected sound pressure level of the *i*-th microphone for the measurement of the transfer source
$L'_{p,\,\mathrm{TS},\,i}$	uncorrected sound pressure level of the *i*-th microphone for the measurement of the transfer source
$\overline{L_{p,\,\mathrm{TS,\,in\,situ}}}$	time and surface averaged sound pressure level of the transfer source in situ
$\overline{L_{p,\,\mathrm{TS,\,no\,scr}}}$	time and surface averaged sound pressure level of the primary source measured without windscreens
$\overline{L_{p,\,\mathrm{TS,\,scr}}}$	time and surface averaged sound pressure level of the transfer source measured with windscreens
$\overline{L_{p,\,\mathrm{unknown}}}$	time and surface averaged sound pressure level of the unknown source
$L_{p,\,\mathrm{uni},\,i}$	sound pressure level of the *i*-th microphone measured with Uniform window
$\overline{L_{p,\,\mathrm{xx}}}$	time and surface averaged sound pressure level of xx microphones

L_W	sound power level
$L_W\left(B_{\text{in situ},i}\right)$	in situ sound power level measured at the i-th atmospheric pressure
$L_W\left(B_{\text{in situ},\max}\right)$	in situ sound power level measured at the maximum atmospheric pressure
$L_{W,\text{cal}}$	sound power level under calibration conditions
$L_{W,\text{dip, free field}}$	free field sound power level of a dipole
$L_{W,\text{dir}}$	sound power level directly determined after sound pressure or intensity measurements
$L_{W,\text{DUT, cal},I}$	sound power level of a device under test under calibration conditions after sound intensity measurements
$L_{W,\text{DUT, cal},p}$	sound power level of a device under test under calibration conditions after sound pressure measurements
$L_{W,\text{DUT, in situ},I}$	sound power level of a device under test after in situ sound intensity measurements
$L_{W,\text{DUT, in situ},p}$	sound power level of a device under test after in situ sound pressure measurements
$L_{W,\text{free field}}$	free field sound power level
$L_{W,\text{in situ}}$	sound power level in situ
$L_{W,\text{known}}$	sound power level of the known source
$L_{W,\text{meridional}}$	sound power level using the meridional path method
$L_{W,\text{mon, free field}}$	free field sound power level of a monopole
$L_{W,\text{PS}}$	sound power level of the transfer source
$L_{W,\text{scan}}$	sound power level using the scanning method
$L_{W,\text{sub}}$	sound power level determined after applying the substitution method
$L_{W,\text{TS, cal}}$	sound power level of the transfer source under calibration conditions
$L_{W,\text{TS, cal},I}$	sound power level of the transfer source under calibration conditions after sound intensity measurements
$L_{W,\text{TS, cal},p}$	sound power level of the transfer source under calibration conditions after sound pressure measurements
$L_{W,\text{TS, in situ}}$	sound power level of the transfer source in situ

$L_{W\text{, unknown}}$	sound power level of the unknown source
L_x	ratio of x-axis displacement over radius
L_z	ratio of z-axis displacement over radius
m	spherical harmonics degree
n	spherical harmonics order
n_{ω}	correction factor for sound power changes due to fan rotation speed variations
$n_{\omega\text{, theo}}$	theoretical correction factor for sound power changes due to fan rotation speed variations
n_B	correction factor for sound power changes due to atmospheric pressure variations
$n_{B\text{, theo}}$	theoretical correction factor for sound power changes due to atmospheric pressure variations
n_{est}	number of estimates
n_{expected}	expected value of the spherical harmonics transform order
n_{max}	maximum spherical harmonics transform order
n_{mic}	number of microphones
n_r	number of measurement radii
n_{rec}	number of receivers
$n_{S\,\text{partial}}$	number of partial measurement surfaces
n_{scr}	sound source radiation order
n_T	correction factor for sound power changes due to ambient temperature variations
$n_{T\text{, theo}}$	theoretical correction factor for sound power changes due to ambient temperature variations
P	receiver
P	sound power
P_{cal}	sound power under calibration conditions
$P_{\text{in situ}}$	in situ sound power
P_n^m	Legendre function

P_{ref}	reference sound power
p	sound pressure
$p(x,y,z)$	sound pressure at a point in cartesian coordinates
$p(k,r,\theta_{\text{el}},\varphi_{\text{az}})$	sound pressure at a point in spherical coordinates as a function of frequency
$p_{\text{dip, free field}}$	dipole free field sound pressure
p_{dir}	direct component of the sound pressure
p_{meas}	measured sound pressure
$p_{\text{mon, free field}}$	monopole free field sound pressure
p_{nm}	sound pressure coefficient
p_{ref}	sound pressure reference value
p_{refl}	reflection component of the sound pressure
$p_{\text{refl, plane}}$	reflection component of the sound pressure for plane waves
$p_{\text{refl, spherical}}$	reflection component of the sound pressure for spherical waves
p_{S_+}	sound pressure of the positive pole of a dipole
p_{S_-}	sound pressure of the negative pole of a dipole
p_{SHT}	sound pressure after the spherical harmonics transform
$p_{\text{SHT},n}$	sound pressure after the n-th order spherical harmonics transform
q	effective complex source strength
$R(\theta)$	angle dependent reflection factor
r	radius
r_i	i-th radius
r_{min}	minimum radius
r_{S}	distance between original source and receiver
$r_{\text{S}'}$	distance between mirror source and receiver
Re	real part of complex number
S	monopole original source

S	measurement surface
$S_{\mathrm{DUT,\,cal}}$	surface for the measurement of a device under test under calibration conditions
$S_{\mathrm{DUT,\,in\ situ}}$	surface for the measurement of a device under test in situ
S_i	*i*-th partial surface
S_{PS}	surface for the measurement of the primary source
$S_{\mathrm{PS,\,ref}}$	reference surface for the measurement of the primary source
$S_{\mathrm{p_1 p_2}}$	cross-spectrum of the intensity probe microphone signals
S_{tot}	total surface area
$S_{\mathrm{TS,\,cal}}$	surface for the measurement of the transfer source under calibration conditions
$S_{\mathrm{TS,\,cal,\,ref}}$	reference surface for the measurement of the transfer source under calibration conditions
$S_{\mathrm{TS,\,in\ situ}}$	surface for the measurement of the transfer source in situ
S_{x}	Measurement surface for the subscripted source
S_0	Reference surface
S'	monopole mirror source
S_+	positive pole of the dipole original source
S_-	negative pole of the dipole original source
$\mathrm{S'}_+$	positive pole of the dipole mirror source
$\mathrm{S'}_-$	negative pole of the dipole mirror source
T	ambient temperature
t	time
T_{cal}	ambient temperature under calibration conditions
T_{dur}	signal duration
$T_{\mathrm{DUT,\,cal}}$	ambient temperature during measurement of a device under test under calibration conditions
$T_{\mathrm{DUT,\,in\ situ}}$	ambient temperature during measurement of a device under test in situ
$T_{\mathrm{in\ situ}}$	ambient temperature in situ

T_{PS}	ambient temperature during measurement of the primary source
$T_{\mathrm{TS,\ cal}}$	ambient temperature during measurement of the transfer source under calibration conditions
$T_{\mathrm{TS,\ in\ situ}}$	ambient temperature during measurement of the transfer source in situ
u	uncertainty
$u(x_i)$	variance of the *i*-th input quantity
$u(y)$	variance of the measurant
u_{n}	normal particle velocity
u_{r}	radial (normal) particle velocity
$u_{\mathrm{r,\ dip,\ free\ field,\ S_+}}$	free field radial (normal) particle velocity of the positive pole of a dipole
$u_{\mathrm{r,\ dip,\ free\ field,\ S_-}}$	free field radial (normal) particle velocity of the negative pole of a dipole
$u_{\mathrm{r,\ dir}}$	radial (normal) particle velocity of the direct sound
$u_{\mathrm{r,\ mon,\ free\ field}}$	monopole free field radial (normal) particle velocity
$u_{\mathrm{r,\ refl}}$	radial (normal) particle velocity of the reflected sound
u_{r}^{*}	complex conjugate of the radial (normal) particle velocity
$u_{\mathrm{S_+}}^{*}$	complex conjugate of particle velocity of the positive pole of a dipole
$u_{\mathrm{S_-}}^{*}$	complex conjugate of particle velocity of the negative pole of a dipole
$u^2(\cdot)$	uncertainty of the quantity in parenthesis
X_i	input quantity
x	x-axis cartesian coordinate
x_0	initial x-axis displacement
x_1	x coordinate of the first point of a line segment
x_2	x coordinate of the second point of a line segment
Y	measurant
Y_n^m	spherical harmonics transform basis functions

y	y-axis cartesian coordinate
y_1	y coordinate of the first point of a line segment
y_2	y coordinate of the second point of a line segment
z	z-axis cartesian coordinate
z_0	initial z-axis displacement
z_1	z coordinate of the first point of a line segment
z_2	z coordinate of the second point of a line segment

List of tables

Bisher erschienene Bände der Reihe

Aachener Beiträge zur Akustik

ISSN 1866-3052
ISSN 2512-6008 (seit Band 28)

1	Malte Kob	Physical Modeling of the Singing Voice ISBN 978-3-89722-997-6 40.50 EUR
2	Martin Klemenz	Die Geräuschqualität bei der Anfahrt elektrischer Schienenfahrzeuge ISBN 978-3-8325-0852-4 40.50 EUR
3	Rainer Thaden	Auralisation in Building Acoustics ISBN 978-3-8325-0895-1 40.50 EUR
4	Michael Makarski	Tools for the Professional Development of Horn Loudspeakers ISBN 978-3-8325-1280-4 40.50 EUR
5	Janina Fels	From Children to Adults: How Binaural Cues and Ear Canal Impedances Grow ISBN 978-3-8325-1855-4 40.50 EUR
6	Tobias Lentz	Binaural Technology for Virtual Reality ISBN 978-3-8325-1935-3 40.50 EUR
7	Christoph Kling	Investigations into damping in building acoustics by use of downscaled models ISBN 978-3-8325-1985-8 37.00 EUR
8	Joao Henrique Diniz Guimaraes	Modelling the dynamic interactions of rolling bearings ISBN 978-3-8325-2010-6 36.50 EUR
9	Andreas Franck	Finite-Elemente-Methoden, Lösungsalgorithmen und Werkzeuge für die akustische Simulationstechnik ISBN 978-3-8325-2313-8 35.50 EUR

10	Sebastian Fingerhuth	Tonalness and consonance of technical sounds ISBN 978-3-8325-2536-1 42.00 EUR
11	Dirk Schröder	Physically Based Real-Time Auralization of Interactive Virtual Environments ISBN 978-3-8325-2458-6 35.00 EUR
12	Marc Aretz	Combined Wave And Ray Based Room Acoustic Simulations Of Small Rooms ISBN 978-3-8325-3242-0 37.00 EUR
13	Bruno Sanches Masiero	Individualized Binaural Technology. Measurement, Equalization and Subjective Evaluation ISBN 978-3-8325-3274-1 36.50 EUR
14	Roman Scharrer	Acoustic Field Analysis in Small Microphone Arrays ISBN 978-3-8325-3453-0 35.00 EUR
15	Matthias Lievens	Structure-borne Sound Sources in Buildings ISBN 978-3-8325-3464-6 33.00 EUR
16	Pascal Dietrich	Uncertainties in Acoustical Transfer Functions. Modeling, Measurement and Derivation of Parameters for Airborne and Structure-borne Sound ISBN 978-3-8325-3551-3 37.00 EUR
17	Elena Shabalina	The Propagation of Low Frequency Sound through an Audience ISBN 978-3-8325-3608-4 37.50 EUR
18	Xun Wang	Model Based Signal Enhancement for Impulse Response Measurement ISBN 978-3-8325-3630-5 34.50 EUR
19	Stefan Feistel	Modeling the Radiation of Modern Sound Reinforcement Systems in High Resolution ISBN 978-3-8325-3710-4 37.00 EUR
20	Frank Wefers	Partitioned convolution algorithms for real-time auralization ISBN 978-3-8325-3943-6 44.50 EUR

21	Renzo Vitale	Perceptual Aspects Of Sound Scattering In Concert Halls ISBN 978-3-8325-3992-4 34.50 EUR
22	Martin Pollow	Directivity Patterns for Room Acoustical Measurements and Simulations ISBN 978-3-8325-4090-6 41.00 EUR
23	Markus Müller-Trapet	Measurement of Surface Reflection Properties. Concepts and Uncertainties ISBN 978-3-8325-4120-0 41.00 EUR
24	Martin Guski	Influences of external error sources on measurements of room acoustic parameters ISBN 978-3-8325-4146-0 46.00 EUR
25	Clemens Nau	Beamforming in modalen Schallfeldern von Fahrzeuginnenräumen ISBN 978-3-8325-4370-9 47.50 EUR
26	Samira Mohamady	Uncertainties of Transfer Path Analysis and Sound Design for Electrical Drives ISBN 978-3-8325-4431-7 45.00 EUR
27	Bernd Philippen	Transfer path analysis based on in-situ measurements for automotive applications ISBN 978-3-8325-4435-5 50.50 EUR
28	Ramona Bomhardt	Anthropometric Individualization of Head-Related Transfer Functions Analysis and Modeling ISBN 978-3-8325-4543-7 35.00 EUR
29	Fanyu Meng	Modeling of Moving Sound Sources Based on Array Measurements ISBN 978-3-8325-4759-2 44.00 EUR
30	Jan-Gerrit Richter	Fast Measurement of Individual Head-Related Transfer Functions ISBN 978-3-8325-4906-0 45.50 EUR
31	Rhoddy A. Viveros Muñoz	Speech perception in complex acoustic environments: Evaluating moving maskers using virtual acoustics ISBN 978-3-8325-4963-3 35.50 EUR
32	Spyros Brezas	Investigation on the dissemination of unit watt in airborne sound and applications ISBN 978-3-8325-4971-8 47.50 EUR

Alle erschienenen Bücher können unter der angegebenen ISBN-Nummer direkt online (http://www.logos-verlag.de) oder per Fax (030 - 42 85 10 92) beim Logos Verlag Berlin bestellt werden.